PUBLIC SAFETY DIVING

PUBLIC SAFETY DIVING

Walt "Butch" Hendrick
Andrea Zaferes
with Craig Nelson

Disclaimer: The recommendations, advice, descriptions, and the methods in this book are presented solely for educational purposes. The author and publisher assume no liability whatsoever for any loss or damage that results from the use of any of the material in this book. Use of the material in this book is solely at the risk of the user.

PennWell Corporation
1421 South Sheridan Road
Tulsa, Oklahoma 74112-6600 USA

800.752.9764
+1.918.831.9421
sales@pennwell.com
www.FireEngineeringBooks.com
www.pennwellbooks.com
www.pennwell.com

Marketing Manager: Julie Simmons
National Account Executive: Francie Halcomb

Director: Mary McGee
Production/Operations Manager: Traci Huntsman
Editor: James J. Bacon
Book and Cover Designer: Brian Firth

Library of Congress Cataloging-in-Publication Data Available on Request

Hendrick, Walt and Andrea Zaferes with Craig Nelson
Public Safety Diving
ISBN13 978-0-91221-294-4
Hendrick, Walt, 1947–
Public safety diving/Walt "Butch" Hendrick, Andrea Zafers with Craig Nelson.
p. cm.
1. Police divers. 2. Search and rescue operations. 3. Underwater crime investigation.
i. Zafares, Andrea, 1965–
II. Nelson, Craig
III. Title
HV8080.D54 H45 2000
363.34'81--dc21 00-025798

Printed in the United States of America

4 5 6 7 8 12 11 10 09 08

Dedication

This book is dedicated to the thousands of skilled, courageous individuals who diligently train to respond to water-related emergencies. When the unexpected occurs, it is they who will be prepared to take action, perhaps saving lives that might otherwise be lost.

Acknowledgments

In submitting this work for publication, the authors acknowledge the help of Craig Nelson in writing the text, Heather Crilley for her office support, Carolyn Pasternak for her work on the tables and charts, and Pete Nawrocky, Fred Curtis, Steve Phillips, and Frank Pagliardi for their photos. We would also like to thank Lt. Kevin Bridgett, Montgomery County Fire and Rescue Department; Monty Leonard, MD, Ph.D., FACEP; Frank Priest; and Lt. Robert McKay, Wantaugh Fire Department, for reviewing this material prior to publication. Finally, we acknowledge the efforts and support of Orlando Abreus; Andy and Diane Alwine; Kenneth Balfrey; Ralph Dodds; Michael Emmerman; Dr. Darlene Esposito; Todd Hanson; David Harrison; Patrick Kilbride; David McCoy; Mike Mulligan; George Saffrowski; Andy Schmidt; Kiurt Semmel; Walt Szulwach; Dan Vircik; and Jason Yates.

Special acknowledgement is given to the following friends and colleagues for contributing their expertise to this book and the field in general: Harold (Andy) Anderson, commercial diving instructor and contractor; William (Bill) L. High, president of PSI, for his knowledge of cylinder safety and for developing worldwide visual inspection programs; Lt. (Navy) David Holland, DCIEM, experimental diving projects officer, for his contributions to naval and public safety divers worldwide on a variety of topics, including critical incident stress debriefing, underwater aircraft crash operations, and equipment testing; David S. Mandel, EMT-P, FDNY, and NFPA-trained hazardous materials medical decontamination technician, for information regarding decontamination procedures and safety; Greg Palomares, retired Army master explosive ordnance disposal technician, deputy chief of the Charles County Dive Rescue Team, and currently employed by the U.S. Domestic Preparedness Program, for information on community emergency preparedness; Edward Rosacker, NAUI and SSI diving instructor and certified equipment repair technician; and Rockie Yardley, technical investigator (crime scene specialist), bomb disposal technician, police instructor, and former explosive disposal and postblast investigation instructor, for his knowledge of underwater explosive recovery operation safety and procedures.

Table of Contents

Preface

Surrounded by impenetrable blackness, you slide your hands and body over the soft mud of the bottom. The only sound is the hiss and gurgle of your regulator, punctuated at intervals by the voice of your tender, giving you directions by way of an underwater communications system. As you push your way through the mud, your hand bumps against a soft object that could be a decayed log. Visualizing it through your fingertips, your mind's eye then tells you that the object you've found is actually an arm and that you've discovered the boy for whom you've been searching.

Taking one brief moment to calm yourself, you alert the personnel on shore. Your tender tells you to surface slowly, and you follow his instructions, carrying the boy with you. On the surface, you inflate your buoyancy compensator device as your tender begins pulling you to the dock. You cradle the young boy in your arms, protecting his face from a light chop and a steady drizzle of rain. As you reach the dock, you help to place him gently onto a waiting backboard covered with blankets. The EMS crew whisks him away, wrapping him further and quickly initiating resuscitative procedures.

As you watch the boy leave, your thoughts are with him. Still, you know that, as a public safety diver, you've already done all you can for him. You've returned him to the world of light and air, and perhaps you've given him another chance at life. Now it's time to get out of the water, take off your gear, and go home. The operation is already a success in that you're out of the water, safe and sound.

Drowning is the third most common cause of accidental death among adults in the United States, and it is the second most common cause of death among children and young adults. Yet the responsibilities of public safety divers go even beyond the exigencies of this statistic. Every day, criminals dump the evidence of their crimes in the water. Often that evidence is the body of a murder victim. It is the public safety diver who must personally enter this alien realm to fulfill an objective. His actions are driven by the needs of the world above.

Public safety diving entails underwater operations with a purpose. Searches are often conducted in black water, where a diver may not even be able to see his own hand in front of his mask. As if such conditions weren't dangerous enough, public safety divers must cope with dead trees, fishing line, and a myriad of other hazards that might snare and entrap him. Whereas sport divers can plan when they want to dive, the public safety diver has no idea when an emergency call will come in. Sport divers tend to dive in good weather, mainly in the warmer months. They can select the type of environment in which they'll operate, and they bear none of the emotional stress of the professional. A sport diver requires only basic gear and basic skills. The public safety diver, on the other hand, may be called to operate in any condition, any time of the day or night, any day of the year. He may know nothing in advance about the site or the water in which he must dive. All too often, that water will be contaminated with chemicals or bacteria. Using solo, tethered-diving techniques, the public safety diver must often deploy rapidly and face an untold amount of stress from various sources, ranging from the media and public pressure to the dispassionate demands of time. Two sport divers using the buddy system can be entirely self-sufficient, yet the professional is dependent on at least four other personnel, and he requires more advanced equipment, knowledge, and training.

All of these factors separate the public safety diver from almost everyone else who ventures underwater, but in one significant respect he is exactly like all the rest: He needs to go home at the end of the day. It is the purpose of this book to help ensure that routine outcome.

CHAPTER 1

The Foundations of a Public Safety Dive Team

Some departments have well-trained, well-equipped public safety divers, while many others depend on teams made up entirely of recreational divers with little or no special training at all. Only a few departments, however, have a properly trained rapid deployment underwater team. If it is your desire to respond to underwater emergencies with the hope of saving lives, recovering evidence, or bringing closure to a grieving family, then you and your department should consider such a team to be the standard.

The first step toward establishing a rescue and recovery dive team is to understand the risks and common problems that many teams now face. For example, most departments don't have well-defined guidelines for these types of operations. Also, the roles of the various responders aren't always entirely clear. For example, what should the fire personnel do? What should the police do? Where should the EMS personnel be? Who is in charge of an underwater incident, the fire department or the police? Too often, department officers don't understand the mission of a team, nor may other public officials who have the power to order these dangerous operations. Thus, the budgets allocated for training and equipment may range from being nonexistent to merely inadequate. Because actual calls tend to be rare, the members of the dive team may not gain half of the real-world experience they should, and EMS and emergency room personnel may not be prepared to manage the victims of long-term drowning and other water-related incidents. Finally, the public may have misinformed expectations about what a dive team is able to accomplish or should be asked to do.

In drawing up a proposal to establish a dive team, you must be able to make specific recommendations about it. The mission that you choose, as well as the overall level of capability that you envision, must be clearly stated. Do you intend the team to be geared for both rescue and recovery, or strictly recovery? Will its capabilities include underwater vehicle extrication, haz mat diving, swift-water diving, ice diving, deep-water diving, and underwater crime scene investigation? Perhaps even more basic is the matter of need. How many lakes, rivers, or quarries are in your area, or how much of your jurisdiction is bordered by open water? How many drownings have occurred over the past ten years, and how many bodies have been recovered? Might any of those lives have been saved by a rescue dive team? One way to answer such questions is to create a survey of recent incidents. Bear in mind that, in compiling statistics, drowning isn't always listed as the direct cause of death, so you may wish to note that such fatalities are often underreported. If you're interested in evidence recovery, you'll need to consult with law enforcement agencies in your area. Once you've gathered the information for your survey, its usually best to place it in table format for inclusion in your proposal.

It's important to specify who will operate the team, whether the fire department, the police, or some other agency. For any number of reasons, including liability, no team should operate in a given area without proper authorization from the agencies within that jurisdiction. Establishing standard operating guidelines from the outset, even if only interim ones during start-up, can help clarify any areas of ambiguity.

FINANCING

The matter of funding is always one of special concern. Hundreds of teams across the country have learned that if you start a team and operate without all the necessary training and equipment, there will be little incentive for the powers that be to award you additional funds to correct the deficiency. From the very beginning, you should state in realistic terms what such a team will cost so that you won't meet with undue resistance later on. Start-up funding for a public safety dive team can run to thirty or forty thousand dollars or more, depending on the size of the team. Besides start-up costs, however, a team must consider the future as well and be sure that funding will be available for at least the next three years. Otherwise, after spending its allocation on start-up costs, a team may find itself high and dry with no ability to use its new gear.

When determining the budget required for a dive team, you'll need to consider the cost of its major components, including equipment, vessels, vehicles, training, salaries, overtime pay, office supplies, insurance, and operating costs. Knowing the requirements of your team will enable you to estimate the amount of funding needed for each category. This in turn will allow you to develop a yearly budget to keep the team in operation.

After determining your budget, you may find that your department is able to fund a dive team, or you may find that there simply isn't enough money available. In many cases, however, you'll fall somewhere between the two. In that event, you'll need to work with the proper authorities to implement a plan that will allow a team to develop and become operational over a few years' time. If you do implement a multiyear plan, be sure that your team doesn't actually become operational until all of the training is complete and the equipment has been purchased. If the taxpayers and town officials see you making do with less, you'll almost certainly receive the short shrift the next time allocations go around.

The cost of outfitting a team properly is considerable. *(Photo courtesy of Frederick E. Curtis.)*

Funding for a public safety dive team can come from a variety of sources, whether private, through volunteer fund-raising campaigns, out of departmental budgets, or directly from the city, county, or state. Such a wide range of possibilities means that your team may be able to draw from several sources, thereby helping you to attain maximum capital for maximum capabilities. Just remember to ensure that all of your contributors are aware of each other so as to avoid the appearance of double-dipping.

Successful fund-raising activities are typically conducted by members of the team themselves and can include raffles, T-shirts, billboards, bumper stickers, and direct-mail advertising. The press and broadcast media can be tremendous allies in your efforts. Invite them to a drill or training session, and ask that they include information in their reports as to how to make donations to support the team.

In many cases, especially in volunteer departments, the team depends on its members to supply some of the personal gear, such as exposure suits, masks, gloves, boots, and fins. Although this can help the team save money, this practice can create liability issues in the event of an injury, fatality, or failed operation. If your team does follow this option, make sure that any of the equipment that the members provide is logged with the department, conforms to department specifications, is properly maintained, and is covered by the department's insurance policy.

Wildcat Teams

In some areas, groups of divers band together to form their own rescue/recovery teams without working under the auspices of a law enforcement agency, fire department, or EMS organization. Although many of these wildcat teams are extremely successful, they often face funding difficulties and legal problems. Such teams must make sure to get the proper authorization to operate as an emergency response unit; otherwise, they could be forced off of the scene and prosecuted for interfering with official emergency responders. Furthermore, because wildcat teams don't have the financial backing of a government entity, they can encounter difficulties attaining insurance and face dire problems in the event of a lawsuit.

Often wildcat teams exist because local fire and rescue departments require that all of their members be certified as firefighters or EMTs, or otherwise possess some skill associated with the department's main mission. Dive team members may only be interested in diving and water rescue, however, and so may lack the proper certification to act as emergency responders. Departments that have such certification requirements may benefit from making exemptions for dive team members. Both sides will benefit from a union of talent and effort. If you have contemplated forming an independent team or are already on a wildcat team, you should look closely into affiliating with a local response unit.

Professionalism

Whether your team is volunteer or paid, the public will view you as professionals. When you respond to a scene, they expect to see an organized, cohesive group in which everyone knows his task and operates efficiently. Equally important to the public's perception is the team's actual appearance. The public expects to see neatly dressed, if not uniformed, personnel responding in official-looking vehicles. Even if your team lacks uniforms and your personnel respond in their own cars, take a few steps to ensure that they make a professional showing.

If possible, team members responding to a scene should don a team shirt, hat, or coveralls. Also, have the personal flotation devices of shore personnel neatly labeled with team names or logos. Even if they don't wear a team uniform, make sure that all personnel are dressed neatly and appropriately for the scene.

Just as you should look professional, keep the site well organized. Don't leave equipment strewn about, park vehicles haphazardly, or allow bystanders to wander through. Avoid horseplay, joking, and standing around with nothing to do. Speak courteously and professionally, and avoid foul language and distasteful humor.

SOPs Versus SOGs

Many teams use only the term *standard operating procedures* to refer to their set of operational rules. Because of a few court cases in which an agency was shown to be derelict in its duties for not following "procedures," many teams are now shifting away from the term SOPs and are opting instead for *standard operating guidelines*, or SOGs. The difference in connotation may be important, at least from a legal standpoint. An action described as a "procedure" implies a rule that must be followed, whereas a "guideline" is more flexible, allowing the responder to tailor his actions to fit the situation.

With this distinction in mind, a team may opt to write both SOPs and SOGs for its operations. The set of standard operating procedures would include any rules that should be inviolable. For example, a diver should end his dive when his air reaches 1,000 psi for an 80-cubic-foot tank. That 1,000-psi limit is a rule that must be not be broken. By way of comparison, a diver should ideally always have his blood pressure taken before a dive. Since this isn't always possible in an emergency situation, any rules governing a diver's blood pressure should fall under the department's guidelines. When writing SOPs and SOGs, use the term "shall" for procedures and "should" for guidelines. This simple difference in terminology will mark the distinction and help you in case of a legal question.

When deciding these issues, simply ask, "What do we have to do?" Anything that must be done belongs in an SOP.

Beyond semantics and legal quibbles, the true nature of either is intended to benefit the course of an operation, establishing minimum requirements and defining roles. Perhaps the operant word in the phrase is neither "procedure" nor "guideline," but "standard." No chief can legally order a firefighter into an inferno or a haz mat situation for which he is unprepared, but that same chief may not think twice about ordering an unprepared diver to enter the water to save a child. This is an event that happens almost on a daily basis around the world. Why? Because there are few written rules, laws, procedures, or guidelines in place governing water rescue and recovery. This is a polite way of saying that there are no true, universally recognized standards.

One of the most important aspects of writing your SOPs and SOGs is in stating what your team can and cannot, or shall not, do. The "shall not do" parts are just as important, if not more so, that the "can do" parts. For example, working in water that is contaminated with hazardous materials or that is moving faster than two knots should both be strictly prohibited. The "shall not do" parts protect both the physical safety and legal liability of the team members.

It should also be explicitly stated that any rescuer, at any point, can say no to performing a dive because of insufficient training, inadequate resources, inordinate danger, or simply because he doesn't feel physically or mentally suited to the task. For any situation that is beyond a team's capability, the SOGs should include the appropriate agency to call. As further protection, your SOGs should define when an operation changes from a rescue to a recovery, and it should employ wording such as "attempt to rescue" rather than "rescue." This sort of terminology can greatly reduce the exposure to liability if a team is unable to save a drowning victim.

Operational protocols, minimum equipment, personnel requirements, qualifications for team membership, and issues of training, drills, health, and safety should all be addressed in your SOGs and SOPs. Review these standards in-house at least annually to see whether any changes are necessary. While doing so, also take the

Procedures and guidelines should state, in specific terms, circumstances under which the team will not operate.

time to review the standards of other teams, and have other teams and trainers review yours.

It is up to you and your department to protect yourselves. Every team member should have a copy of the rules you write, and each should sign a statement indicating that he has read, understands, and agrees to abide by them.

OSHA Regulations

The United States Department of Labor's Occupational Safety and Health Administration (OSHA) sets forth commercial diving standards in Standard No. 1910.401 (a). These regulations, however, do not apply to public safety diving, as stated in OSHA guidelines. In fact, the agency maintains that public safety diving operations are exempt from strict OSHA standards altogether. It should be noted, however, that as long as the depth of an operation does not exceed the no-decompression limit, or a depth of one hundred feet of seawater, dives that conform to the principles expressed in this book do, as realistically as possible, meet OSHA regulations for scuba diving operations.

Although dives conducted for public safety purposes by or under the control of a governmental agency are exempt from OSHA standards, teams should note that other OSHA regulations may apply to other types of diving. Ice diving, submerged vehicle penetration, cave diving, and other types of overhead environment diving may fall under OSHA regulations for confined spaces, generally defined as areas with only one route in and out. OSHA regulations may apply to mixed-gas diving as well. Because safe techniques for these operations aren't fully addressed in this book, the authors strongly advise that teams first consult OSHA regulations and legal experts before attempting such dives.

Because OSHA regulations were developed to keep commercial divers hale and hearty, they are well worth the time it takes to review them. The commercial diving industry, for which and by whom those regulations were developed, has far more diving experience than the public safety diving community, and their expertise should not be overlooked.

Insurance and Liability

In a society in which lawsuits are rife and medical costs are high, it's obvious that dive teams need insurance. A team may be held liable for failure to accomplish a rescue, or for a rescue or recovery that it performs improperly, and any of these scenarios may occur for a variety of reasons. Even the best of teams can be accused of negligence or taking inadequate action to save a life. Although one might think that a volunteer dive team would at least be safe from litigation, the courts have proved otherwise. There are no longer any volunteers in rescue/recovery, only professionals, with the difference being that some are paid and some are not. Being considered a professional leaves anyone open to litigation, and exposure to litigation demands insurance.

Bear in mind that the origin of a lawsuit might not necessarily be outside the team, for the team could be sued by one of its members or a member's family in the event of a death or injury. Also, a team might be sued for damage that it inflicts on private property during the course of an operation. When talking to insurance agents and department lawyers, be sure that your team has coverage that is adequate for the potentialities. If your rescue department already has insurance, make sure that the coverage extends to underwater operations. Remember, too, that teams should keep all records from training and actual calls for a minimum of five years after an event, but keeping them for seven years is recommended. Having records could be essential in case of a lawsuit or investigation, either of which may be initiated years after the event itself.

Individual Insurance

For the individual team member, the subject of insurance hinges on whether diving for the team is part of his job description or whether he is a volunteer for the team. For team members who are paid professionals, the matter of insurance is a fairly simple one, falling under workers' compensation. For volunteers, the issue is a bit more complex.

Volunteers should receive some type of compensation for injuries sustained while diving for their department. In addition, volunteers should have some type of personal health insurance that will cover at least nondiving injuries sustained while working for the dive team and that will also provide coverage for vaccinations, especially for tetanus and hepatitis. Volunteers should also carry some type of diving insurance. This type of insurance is secondary, covering in-water injuries and conditions, such as lacerations from sharp debris, embolisms, and decompression illness. The Divers Alert Network (DAN) Master Plan is the most suitable for public safety divers. Finally, volunteers may also wish to have volunteer coverage to protect against litigation. This insurance may be an option on a homeowner's policy, or it can be a separate liability policy. Such coverage is easily available and generally inexpensive if membership on the dive team isn't part of your regular profession. Check with your department about the level of coverage you need.

Property Insurance

As a paid member of a dive team, the equipment that you use is generally covered by the department's policy. For volunteers, however, coverage is unlikely to extend to personal equipment unless it is recorded as part of the dive team's regular operating equipment. If the equipment isn't listed with the department and you are injured or killed, department lawyers can attempt to show that your gear was at fault. If it is proved that your gear was at fault and it wasn't listed with the department, you may lose your workers' compensation or your family may not receive death benefits. In addition to listing it with the department, be sure that your equipment is also listed on your homeowner's or renter's insurance policy.

Study Questions

1. True or false: In drawing up a proposal to establish a dive team, you must be able to clearly state the level of capability that you envision.

2. Besides start-up costs, a team must consider the future as well and be sure that funding will be available for at least the next __________.

3. What is a wildcat team?

4. An action described as a ________ implies a rule that must be followed, whereas a ________ is more flexible, allowing the responder to tailor his actions to fit the situation.

5. May a public safety diver refuse to go underwater simply because he doesn't feel mentally suited to the task?

6. Do commercial diving operations fall under OSHA regulations?

7. True or false: OSHA maintains that public safety diving operations are exempt from strict OSHA standards altogether.

8. True or false: OSHA regulations may apply to underwater confined-space operations, as well as mixed-gas dives.

9. For legal purposes, a team should keep all records from training and actual calls for a minimum of ________ years after the event, but keeping them for ________ years is recommended.

10. For volunteers, insurance coverage is unlikely to extend to personal equipment unless you take what action?

Chapter 2

Dive Team Command

Although a dive team may have all the expertise and equipment in the world, it may not be up to the challenge of executing a successful dive operation. A team can do little without solid leadership. A good team leader can make the difference between a rapid, successful save and a sloppily executed body recovery.

Let it be said early on that individual members may need to wear many hats. The average dive team responds with a total of five to seven members, requiring each surface support member to be assigned two or more roles. The one benefit of having a small number of responders, however, is that fewer officers are needed.

The control of a team in operation should be based on the incident management system (IMS), otherwise known as the incident command system. IMS organization and terminology differ between departments, and aspects of it may even vary within a given department, depending on the type of incident at hand. The version described below is based on an IMS structure suitable for haz mat responses, since water is essentially an alien environment that can kill. The command system described in this chapter can be altered to suit larger and smaller teams.

Off-Site Command

Truly solid command begins before the dive truck ever leaves its headquarters. Because public safety diving requires so much support and preparation, a team needs good leadership before it ever becomes operational, and that leadership comes from off-site command.

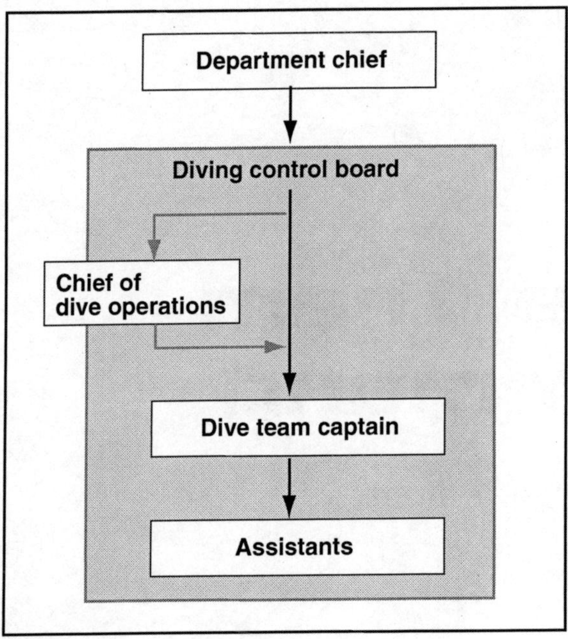

Off-site command structure.

Within a department's command structure, a dive team may be headed by the team's captain, who must be an experienced dive team member. The dive team captain may also appoint assistants, and these should be experienced divers and tenders.

A diving control board consisting of the dive team captain and his assistants should be established to make decisions about operations and other matters, such as standards of operation, budgeting, and equipment purchases. Large teams may wish to add a chief of dive operations at the top level to act as an overall liaison to the department chief.

On-Site Command

At the scene of an incident, there must be only one commander and one command post, thus creating a unified command for coordinating the efforts of all the agencies involved. This unification is especially important for water-related incidents, where the duties and functions of different agencies are typically less recognized than they are at other types of emergencies. Too often, the police, firefighters, EMS personnel, and bystanders all wind up in the water, searching for the victim without an effective plan. Such a situation is a prelude to disaster. Perhaps a police dive team and a fire department rescue squad arrive simultaneously on the scene. Which agency should handle the operation, and how can they work together for optimal results?

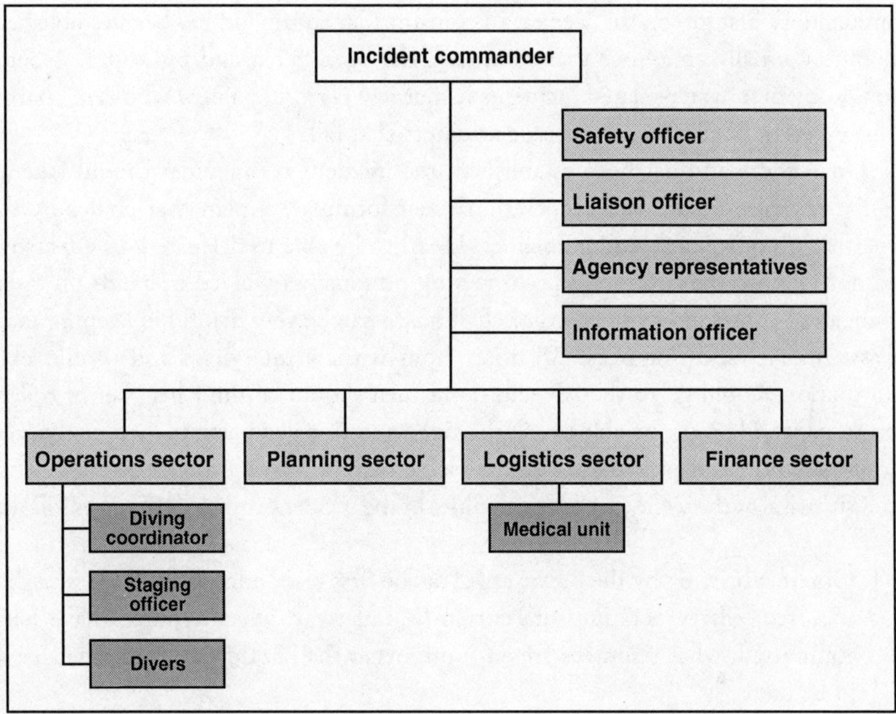

On-site command structure.

Turf and jurisdictional wars shouldn't be resolved during an actual incident. Issues of authority should be decided long before the alarm ever sounds. One county in New Hampshire has implemented a rule that, as long as the incident is in the rescue mode, which includes up to an hour of submergence without it being a known crime scene, then the fire department has jurisdiction. Once the incident moves into a recovery mode, then the police are in charge. Whether you adopt this model or not, you should take the time to work out such issues with the other agencies in your area, well in advance of the next call.

Incident Commander

Traditional rank and titles mean little when it comes to who should be the incident commander (IC). The first-arriving officer usually assumes that role until passing it off to a senior officer or a peer with more experience. An incident commander must be observant. He should have solid leadership skills, and he should be certified at least to the Awareness level of training for diving operations. (The Awareness level is the minimum required for public safety dive operations, usually consisting of several hours of training on such topics as safety and securing the site.) The incident commander's first job on the scene is to confirm that command has been established. He must formally announce that command has been taken and by whom. A common problem in water-related incidents is that few personnel know who is in charge. This results in freelancing and rescue attempts that fail.

Once command has been established, the incident commander should take an omniscient view of the entire operation, then formulate a plan that abides by the department's procedures and guidelines. He must be able to delegate jobs effectively and not fall into the common trap of getting personally involved in hands-on tasks. An incident command whiteboard or chalkboard can be very helpful in keeping track of assignments and resources. All notes from witness interviews and profile map information should go to the IC, and he in turn should commit his plan of operation to paper. He must be able to ask the right questions and assimilate the advice of various subordinates and advisors into a viable plan of operation.

On arrival at the scene, the responsibilities of the incident commander are as follows.

1. Obtain a briefing by the previous IC or the first responder to find out what has occurred, what operations are currently underway, what resources have been committed, what resources are en route, what the hazards are, and what needs to be done.
2. Assess the status of the incident, then establish objectives and projected resource requirements.
3. Give an initial briefing and set up an IMS structure.
4. Assign a safety officer, liaison officer, dive coordinator, and other officers and personnel as required.
5. Authorize a plan of action and decide whether the operation is a go or a no go, based on a risk-to-benefit analysis.
6. Manage the command staff.
7. Approve requests for resources.
8. Approve the release of information to the media by the liaison officer.
9. Document the incident.
10. Approve plans for demobilization.

Sometimes too few personnel respond to the scene. Other times, far too many agencies and personnel are present. The IC is equally important in both situations, but he'll need to handle them differently. For example, if too few divers are available to conduct a safe operation, the IC must ensure that no one enters the water until the minimums are met. If too many divers are on the scene, the IC must ensure that only the requisite number suit up. As the number of personnel increases, the more assignments the IC should delegate. Whatever he doesn't delegate is his own responsibility.

An effective commander knows not to rush an operation and to slow down other responders who do. The movements of a dive team must be disciplined and professional, meaning that every action should have a planned purpose. Rushing typically results in mistakes, injury, and broken equipment, and it usually ends up wasting precious time. Ultimately, it is up to the incident commander to ensure that all personnel perform the right actions, in the right way, in the right amount of time.

If the IC is replaced for any reason, it's up to the new commander to decide whether the operation is going the way it should. The new IC shouldn't be afraid to change the operation if it is unsound. Prolonging an unsafe or infeasible operation will only have repercussive effects as the incident progresses.

The IC, as well as all the other officers of the command hierarchy, must be easily recognizable. Color-coded IMS vests can help identify them. This is especially important if the role of incident commander is passed off several times.

Diving operations must be performed correctly, with the proper equipment, procedures, and trained personnel. If any component isn't ready and in place, a good incident commander will shut down the operation until it's safe to proceed. If the situation is beyond the means of the available resources, then the operation should halt, pending the response of the appropriate agencies. In an emergency situation, where lives are often at stake, such decisions aren't always easy to make. It's up to the incident commander to make the call.

Command Post

The command post should be easily accessible, readily apparent to all involved in the incident, and usable in difficult weather. It must be staffed at all times. All information pertaining to the incident should be compiled there. The IC should have a visual master map of all the different sectors of the operation, and he should receive regular updates on their status. The parameters of an emergency may quickly change. Proper documentation will make it easier to adapt strategies and tactics to meet those changes.

Command Unit

Dive teams often become so focused on their dive trucks that they forget one of the most important vehicles on any site: the command vehicle. Although many teams can't afford the luxury of a dedicated command vehicle, it's important that some type of mobile unit be viable as a command post. That unit could be a truck, trailer, security car, fire chief's buggy, or even a box that can be transferred from one vehicle to another. As long as the unit contains everything the command post needs, it doesn't matter what type of package it comes in.

A command unit should be equipped with a radio capable of all frequencies that may respond; two cellular telephones (one for incoming calls and one for outgoing calls); road maps of the area; phone books; emergency and mutual aid plans; pens, paper, clipboards, whiteboard, markers, and easel; duct tape and tie wraps; binoculars;

evidence preservation containers and tags; command vests; rain gear; personal flotation devices with suitable ancillary equipment; spare warm clothing; portable radio chargers; a multiport cigarette lighter adapter; and appropriately colored lights for your agency. Your team might also consider equipping the command unit with a few simple rescue items, such as rescue rope throw bags, first aid kits, blankets, and a backboard.

Zones

The scene of a water-related incident can be divided into three zones: hot, warm, and cold, as in a haz mat operation. Command must delegate personnel to set up and enforce the demarcation of these zones. The hot zone is the water, reserved for Technician-level personnel (the highest level of training) deployed to effect the rescue or recovery. The warm zone is where Operational-level personnel (the middle level of training) conduct the onshore support procedures. The cold zone, the shore area behind the warm zone, is for the media, family members, and bystanders. Typically, in diving operations, the whole world seems to respond, yet only a handful of the individuals who turn out will be qualified for water operations. Hence, the command post is often initially founded in the warm zone. As additional qualified personnel arrive, the command post can move back into the cold zone, where it will have a better vantage of the entire operation. If space permits, an EMS rehab staging area should be established in the warm zone for pre- and postdive checks of dive personnel, although EMS personnel for the victims may best be stationed in the cold zone until called into play. The haz mat decontamination area should also be in the warm zone, set up near the access point to the hot zone.

For reasons of accountability, each zone should have only a single access point. The IC should appoint someone to man these points of entry and keep track of who goes in and out. At any time, the officers in the command post should know how many people are operating within each zone.

The boundaries between the zones should be marked and taped off. Some teams use red tape to demarcate the hot zone. The hot zone usually begins at the shoreline; however, depending on the circumstances, Command may decide to place the boundary of the hot zone farther inland to protect Operational-level personnel. The size of the warm and cold zones will be dictated by the operation, the number of qualified responders, the amount of equipment, and other factors. Remember, however, that it is better to have zones that are too large rather than too small. You can always make a given area smaller, but it can be very difficult to push back a crowd of bystanders to gain more room for response vehicles.

Manageable Span of Control

Although the incident itself may not require many members, it may take extra officers to manage the number of personnel and agencies that do respond.

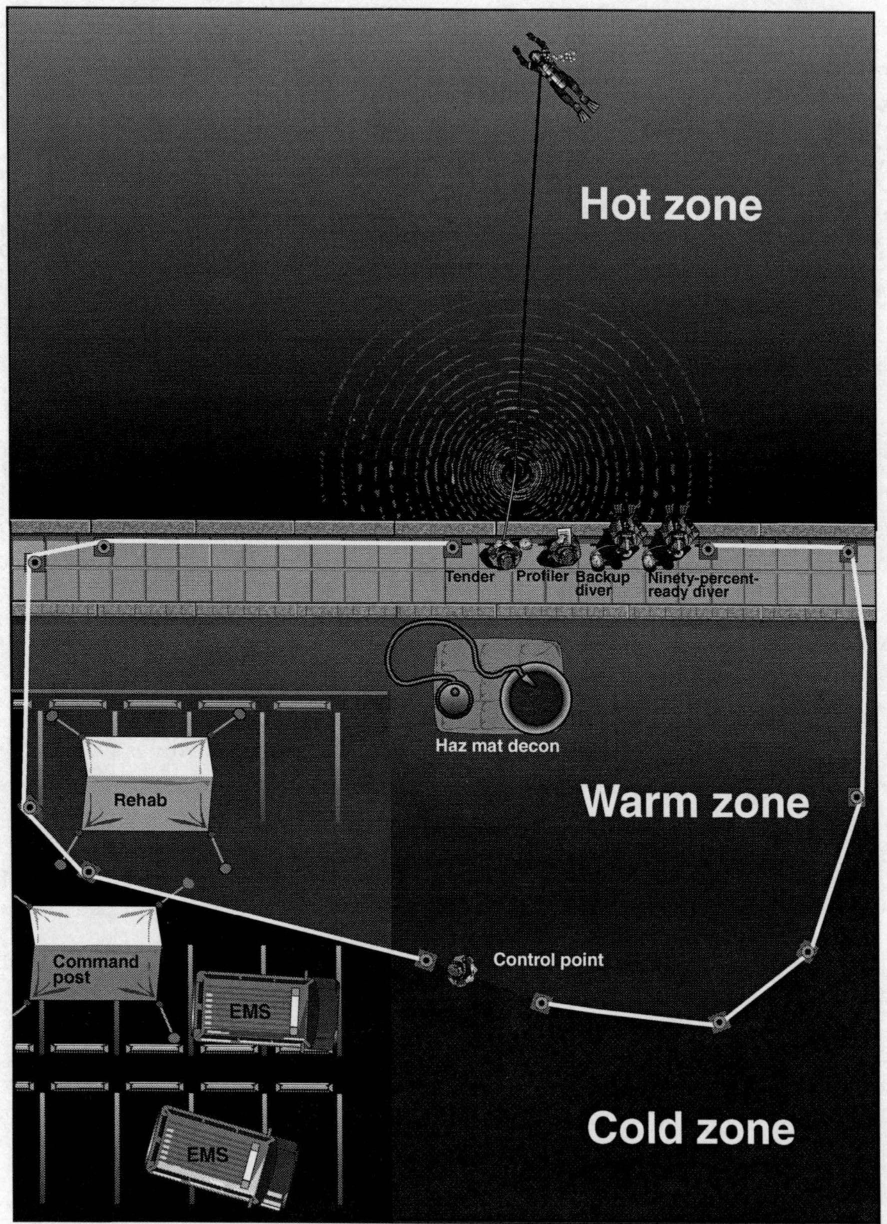

The scene of a water-related incident can be divided into three zones.

Maintaining a manageable span of control means delegating authority to a sufficient number of subcommanders, who in turn ensure that all of the personnel under them are assigned duties and are following them properly. From a management standpoint, one leader or supervisor for every three to seven subordinates is optimal. Having a manageable span of control is especially important for water-related incidents because, as discussed earlier, the tasks are not routine, the equipment may rarely be used, and few personnel may be sufficiently familiar with the department's procedures and guidelines.

While retaining overall command, the IC should choose sector officers to oversee the various aspects of the incident. These officers should report regularly to the command post in person or by radio. Like the IC, the sector officers must maintain a view of their entire area of operation and not become focused on minutia.

Safety Officer

The incident commander must always designate a safety officer to identify, assess, and correct hazardous conditions and unsafe practices. This officer must oversee the welfare of all personnel on the scene. Most importantly, he has the authority to stop any operation that he deems imminently hazardous. Ideally, the safety officer will be certified to the Technician level of training. For a large operation or when a crowd of bystanders is present, the incident commander may elect to designate a safety officer for the overall operations and a separate Technician-level dive safety officer to oversee underwater operations only.

Among his various duties, the safety officer should obtain briefings from the IC, participate in planning at the command post, review the plans for the operation, continuously assess the safety of the operation, investigate any on-site accidents that occur, and document any information pertinent to his sector.

The responsibilities of the safety officer are as follows.

1. Don personal protective equipment (PPE), plus an IMS vest or some other identifying item of clothing.
2. Determine and mark the zones of operation, if not already done.
3. Ensure that no one enters the warm zone without authorization and the necessary personal protective equipment.
4. Request law enforcement personnel to help with crowd control, if necessary.
5. Ensure that any medical needs of the witnesses are being met.
6. If weather conditions warrant, ensure that medical personnel are waiting inside the ambulance rather than getting cold during the search effort.
7. Ensure that medical personnel have appropriate transport equipment ready to ferry the victim from the shoreline to the ambulance, and that the medical personnel wear personal flotation devices if they have to approach the shoreline.
8. Ensure that the access routes to the shoreline and the ambulance are free from obstructions and hazards.
9. Ensure that the blood pressure of each diver is taken before dressing and that each diver has a tender to help him dress.

10. Ensure that no diver enters the hot zone without the proper equipment, including a pony bottle, harness, tether, and a weight belt with an easily accessible release.
11. Ensure that every tender is wearing a fully zippered personal flotation device, gloves, and other necessary equipment.
12. Enforce all safety procedures and guidelines.
13. Ensure that all tenders stay by their divers until the divers are dressed down and checked by EMS personnel.
14. Ensure that no personnel run or rush, increasing the chance of injury and escalating stress levels.
15. Shut down any unsafe procedures or actions immediately.

Dive Coordinator

The dive coordinator (DC) functions as one of the sector officers in the incident management system, supervising the overall diving operation and working closely with the safety officer to ensure a nominal outcome. This officer doesn't enter the water, but rather, should maintain a position that affords him a good overall view of the scene. If there are more than five operating dive teams or if the incident is spread out, the DC should designate assistant dive coordinators.

The safety of the dive team personnel is the primary responsibility of the dive coordinator. He should evaluate the site and consider any factors that may affect the operation. Thus, the first job of the DC at every scene is to conduct a risk-to-benefit analysis, then report the pros and cons to the incident commander, including whether the operation should be undertaken immediately, later, or never. Taking various factors into consideration, the DC should also decide whether the operation should proceed in a rescue or a recovery mode. In requesting divers to conduct underwater operations, the dive coordinator should never ask his personnel to perform any task that requires a certification that they do not possess, and he should carefully evaluate any new members to ensure that they have the experience, qualifications, and savvy for a given task. The DC should also evaluate blood pressures, attitudes, and the overall appearance of divers to help make safe deployment decisions. Any hesitation by a diver should be interpreted as a desire not to go underwater. No member should ever be coerced into making a dive that he doesn't want to make.

As the search progresses, the DC needs to stay informed about the rates of air consumption, since excessive consumption may indicate an equipment problem, fatigue, or an inexperienced diver. He should also ensure that a profile map of the diver's exact movements and the entire search area is made by the backup tender, and he should closely monitor the progress of the operation. Frequent slack in the line or the inability of a tender to follow correct procedures may mean that a given area has to be searched again, and a large area that has been searched too quickly may not have been checked thoroughly.

The safety of dive team personnel is the primary responsibility of the dive coordinator.

Public Information Officer

A team essentially has two options in dealing with the press. The first is to ignore reporters and completely restrict them from the area. This is usually a bad option, since it fosters the impression that the dive team is working with no information and no plan. The better option is to keep the press informed of the operation and the plan by which it is proceeding. It is the responsibility of the public information officer to provide that interface between the team and the press. As a communicator, the PIO should be able to explain the hazards of public safety diving so as to provide an understanding of why such operations can take so long and require so many precautions. If the PIO can describe the complexity and safety issues of the mission, he will win the positive sentiment of the press and the public. In fulfilling his duties, however, he must also ensure that the family members of victims are always given information before passing it on to the press. That way, no one in the family will find out by radio or TV that a loved one has been found.

By working with the members of the press rather than treating them as the enemy, dive teams can gain much-needed public support. Still, neither the IC nor the operation as a whole should be hindered in any way by the press, family members, or bystanders, so the PIO also acts as a buffer between these groups. Like the safety officer, the PIO helps to keep those who aren't directly involved in the operation in the cold zone, a task that may require the assistance of law enforcement personnel.

The essential responsibilities of the public information officer are as follows.

1. Obtain briefings from the incident commander.
2. Assess the incident and summarize its significant facts and points of interest.
3. Obtain approval from the IC to release information to the press.
4. Arrange meetings between the IC and the press.
5. Keep the press and other bystanders in the cold zone.
6. Fill out the appropriate documentation.

Liaison Officer

It is imperative that all of the agencies on the scene act in a coordinated manner, so for a large operation, it may become necessary to appoint a liaison officer. This officer serves as a link between the IC and other agencies. In a practical sense, he serves as a diplomat toward agencies that aren't accustomed to working within the incident management system. He determines where each agency can stage, ensures that they have adequate two-way communication, and prevents other entities from setting up their own command posts.

The principal duties of the liaison officer are as follows.

1. Obtain briefings from the IC.
2. Locate each agency's representative and establish two-way communication and coordination.
3. Obtain additional outside resources as requested by the IC.
4. Fulfill the needs of other agencies as requested by their representatives.
5. Monitor the operation to discover, prevent, and cure any interagency problems.
6. Fill out the appropriate documentation.

Staging Officer

By definition, a staging area is a place where personnel and resources are stationed to wait for their assignments; where personnel ready themselves and their equipment; and where personnel later dress down and repack their equipment. Designating such an area for responders serves both accountability and management. Each agency may have its own staging area, and if the operation is large enough, each staging area will be controlled by a staging officer, who is an arm of the incident commander. In a small operation, the duties of the staging officer would be covered by the diving coordinator and the safety officer.

The basic responsibilities of a staging officer are as follows.

1. Obtain briefings from the IC or the liaison officer.
2. Set up the designated staging area.
3. Establish check-in and checkout accountability procedures for personnel.
4. Dispatch resources when requested by the IC.

Agency Representative

A representative from each responding agency should check in with the incident commander or liaison officer at the command post, participate in planning, and inform the liaison officer of any special needs of his particular agency. For smaller operations, the job of the agency representative could be covered by a staging officer or the diving coordinator.

Medical Unit Leader

The medical unit leader makes sure that there are enough warm EMS personnel in the ambulance, as well as enough to look after the divers, support personnel, and potentially anyone else on the scene. In short, this officer must be ready to address the medical needs engendered by the entire operation.

The primary responsibilities of the medical unit leader are as follows.

1. Select a triage officer, in the event of a mass-casualty incident.
2. Obtain briefings from the IC or liaison officer.
3. Determine, from a medical standpoint, the current status of the incident and what needs to be done.
4. Determine whether sufficient resources are on hand or available, including EMS personnel who have a basic knowledge of diving emergencies.

5. Prepare a medical emergency plan, including objectives and procedures.
6. Ensure that the medical emergency plan is carried out.
7. Respond to requests for first aid by personnel, family members, bystanders, and victims.
8. Monitor personnel before, during, and after the operation to ensure that they stay sufficiently warm and hydrated.
9. Ensure that the blood pressures of divers are checked both before and after each dive.
10. Prepare documentation and a medical report.

Technical Specialist

A technical specialist is someone who may be from outside the public safety industry but who can provide specific expertise needed at the scene. Suppose, for example, that a recovery requires a dive to 165 feet. Normally, public safety divers don't belong deeper than sixty feet unless they have been appropriately trained and equipped. Therefore, a fully certified, licensed commercial diving company may need to be brought in to make this recovery.

Knowing the right technical specialists to call is imperative toward preventing injury, damage, and a waste of finances. An all-too-common situation involves recovering a submerged vehicle. You may have to find a certified commercial diving company that can demonstrate the proper training, equipment, experience, and procedures to perform this job safely. Other examples of technical specialists relevant to underwater recovery include surface-supplied divers, moving water specialists, contaminated water specialists, and underwater crime scene specialists.

Postincident Debriefing

Postincident analysis is one of the most important aspects of the incident management system. After an operation ends, the incident commander and all of the officers should review what occurred to identify the strengths and weaknesses of the response, then develop any necessary corrective measures to meet the next emergency. In a formal debriefing, the questions to ask should include the following.

1. Were the department's SOPs and SOGs followed, and if not, why?
2. Do the procedures and guidelines need amending? Note that checking to see whether they were followed is an excellent way to determine whether they need revising.
3. Were there enough or too few personnel?

4. Did all of the responding personnel have sufficient training?
5. Were sufficient resources present?
6. Were there any equipment failures or damage?
7. Was the incident properly documented? The records should be reviewed during the debriefing.
8. Did all of the responders behave appropriately?
9. Was all of the equipment properly cleaned and stored for later use?
10. Were all of the divers checked out by EMS, and are they physically and emotionally well?

Study Questions

1. The incident management system described in this book is based on an IMS structure suitable for ________ responses.

2. The means by which the efforts of all responding agencies are coordinated is known as __________.

3. An incident commander should be certified to at least what level of training for diving operations?

4. Where should all information pertaining to the incident be compiled?

5. Who may enter the warm zone of an underwater operation?

6. True or false: Officers in the command post should know at all times how many people are operating in the hot and warm zones.

7. An optimal span of control is to have one leader or supervisor for every ________ to ________ subordinates.

8. For diving operations, the safety officer should ideally be certified to what level of training?

9. When should the dive coordinator enter the water?

10. True or false: A dive coordinator should interpret any hesitation on the part of a diver as a desire not to go underwater.

11. True or false: In fulfilling his duties, the public information officer must always ensure that the family members of victims are given information before passing it on to the press.

12. The officer who serves as a link between the incident commander and other agencies is known as the ________.

13. In small operations, the duties of the staging officer would be covered by what two entities?

14. List some of the questions to ask during a postincident debriefing.

Chapter 3

Dive Team Personnel

Since diving and water-related calls often end up with fewer qualified responders than other types of incidents, reliability is a basic criterion for selecting team members. Underwater work isn't only physically demanding and potentially dangerous, it's also mentally taxing, even when the efforts prove successful. Recovery operations bring on their own stresses, such as those associated with the handling of decomposed bodies or dangerous materials. The safety of the divers rests largely on the skill and dependability of their tenders, and the success or failure of an entire mission may very well hang on the performance of the individual.

A reliable team member is one who actively participates in drills and actual operations, never shortcutting procedures. Thoroughly familiar with the department's mandates, he works well as a team player and doesn't seek just the high-profile jobs. Rather, he pitches in for the lowly tasks, such as cleanup, routine maintenance, and documentation. At all times, he shows active concern for his peers, the operation, and the public, as well as himself.

Two general aspects of teamwork are required of the various members. First, the members must be able to work with one another. Second, they must be able to work with other teams and associations.

Both surface and in-water personnel should be capable of internal teamwork; that is, working together in a unified way. Such coordination of effort requires training, practice, patience, and strong communication within the team. The members must respect each others' talents and abilities. A team member should be able to recognize when a more experienced person is better suited to a given task. Conversely, the more advanced team members should know when to give others a chance to gain the experience they need.

Interagency teamwork is equally vital to an operation. Too often, various teams and departments battle one another over how different calls should be handled. Such teams often lose sight of the basic goal of making recoveries and saving lives. There is little excuse for squabbling matches and power struggles, for these only serve to undermine budgets, morale, and community relations.

True professionals, whether paid or volunteer, are capable of working together smoothly to accomplish a given objective. Sometimes when state-paid professionals arrive on the scene, the local volunteers are pushed aside. This may not happen at every incident, but it does occur. To avoid this, well-trained volunteers must seek to gain the recognition and respect that they deserve. This is best done through meaningful communication with other agencies, before the incident occurs. If you are on a volunteer team working an operation when other teams arrive, you should not feel threatened. A representative for your group should be able to demonstrate that you are organized and in control of the situation. He should foster a working relationship with the captain of the other team and not hesitate to provide pertinent information, such as profile maps and the credentials of the volunteers.

Because of the type of work that they do, state-sponsored dive teams are trained to take over command of a scene automatically. Some state statutes actually require that the moment a police officer is on the scene, he must take control, whether or not he is trained in the needs of that operation. Still, the volunteers should not be tossed aside. In fact, a display of organization and internal teamwork are more important at that point than at any other.

Health Requirements

The demands placed on public safety divers are high. These members must be able to swim against strong currents, hold on to bodies, disentangle themselves from almost any kind of snare imaginable, and handle heavy gear. Such requirements leave little room for those who swim poorly, are overweight, or who suffer from health problems.

Basic Qualifications

Diver candidates with known psychological problems or a history of claustrophobia should be watched carefully and eliminated from training if there is any doubt in the mind of the instructor as to their abilities. Similarly, personnel who don't wish to serve as divers should never be forced to receive dive training. They can perhaps be trained as tenders instead. If a person doesn't want a job, he won't be able to perform it safely or effectively.

Medical Examination

To be certified as an open-water diver and for membership on a team, all divers must pass a physical examination. Each candidate should also obtain a letter of approval from the examining physician, stating that he is medically qualified to pursue rigorous diving activities. The candidate should keep a copy of that letter in a personal folder, aside from the copy to be kept at the team's headquarters. Preferably, the examining physician should be one who is knowledgeable in diving medicine. Divers should also have a medical exam annually; after any major illness, injury, or surgery; as well as after any diving-related accident. Any number of medical conditions may contraindicate certification as a public safety diver, including ear disorders, inadequate visual acuity, neurological problems, heart condition, hypertension, anemia, asthma, diabetes, hernia, pregnancy, physical deformities, mental irregularities, to name a few general categories. An updated listing of these and more specific contraindications is available from the Divers Alert Network.

Vaccinations

Both divers and surface personnel may frequently be called on to operate in biologically contaminated water, and they should therefore have a current hepatitis vaccination even before being allowed to train with the team. Additionally, because of the plethora of jagged debris in and near the water, all dive team members should be required to have a current tetanus vaccination. A polio vaccination is also recommended.

Swimming Skills

Public safety divers often deal with adverse water conditions. For their own safety, all divers and even tenders should be tested for swimming skills before being allowed onto the team. It's also advisable to conduct annual swimming tests to ensure that the members continually perform to standards. As a practical minimum, a diver should be able to perform the following.

1. Swim nonstop for 200 meters in under six minutes. The next goal would be to swim nonstop for 1,000 feet within ten minutes without swim aids. Any head-forward stroke or combination of strokes is allowable for either test.
2. Swim underwater without swim aids for a distance of 75 feet without surfacing.
3. Swim underwater without swim aids for a distance of 150 feet, surfacing no more than four times and taking only a single breath each time.
4. Without swim aids, tread water for ten minutes or for two minutes with his hands on top of his head.
5. Without swim aids, transport another person of equal size for a distance of 75 feet.
6. Tow an unconscious person nonstop for 150 yards while wearing full dive gear.
7. Hold his breath to retrieve a ten-pound object from the bottom and hold it on

the surface while treading water for at least fifteen seconds, with a goal of reaching over one minute.

Depending on the type of operations anticipated, higher standards may be necessary for divers. Because they work around the water, tenders and other shore personnel should also meet minimum standards. All tenders and support personnel should be able to do the following.

1. Swim for 100 yards while wearing a zipped personal flotation device.
2. Float for ten minutes while wearing a zipped PFD.

If your team has no swimming requirements for tenders, you must stress the use of PFDs even more, since you may unknowingly have nonswimmers on the team. Since even a PFD is no guarantee against drowning, you may wish to institute minimum swim standards anyway.

As an additional measure of physical fitness and ability, it is recommended that divers be able to perform the following.

1. Thirty-five sit-ups in ninety seconds.
2. Twenty-five push-ups.
3. Three chin-ups.
4. Run for one mile in under twelve minutes.

The dive team captain should continuously evaluate the physical and mental condition of his personnel and remove from duty any who are obviously less than fit to remain active as divers.

Diving Skills

Before being allowed to dive with the team, certified divers should be asked to demonstrate their skills in a pool. A diver should be able to do the following.

1. Assemble and don dive gear with the assistance of a tender.
2. Interpret hand and line signals.
3. Remove his mask underwater, then remove and replace his shears, remove and replace his fins, and continue to breathe for at least two minutes without a mask, thereby demonstrating his ability to function devoid of a mask.
4. Doff and don the buoyancy compensator device underwater.
5. Switch to a pony regulator and make an ascent without a mask.
6. Establish and maintain neutral buoyancy.

7. Keep the line taut while performing an arc search pattern.
8. Doff and don the weight belt underwater.

Levels of Certification

As promulgated by the National Fire Protection Association (NFPA), there are three levels of certification for those who respond to water-related emergencies: Awareness, Operational, and Technician. These distinctions apply to all members of water operations teams, including law enforcement personnel and EMS crews. The immediate duties of each level are described below and discussed in detail in later chapters.

Awareness Level of Training

Anyone who would respond to a water rescue or recovery should be trained at least to the Awareness level, which involves at least four hours of training. This first level of training has the important main goal of preventing further injuries and problems at the scene.

On arrival at the scene of a water-related emergency, Awareness-level personnel should do the following.

1. Don a personal flotation device and other necessary personal protective equipment.
2. Determine where to stage the operation.
3. Assess the scene to determine what sort of response is necessary. The responder should determine the number of victims and mark a spot on shore in front of each one, if possible. He should also identify any hazards that might pose a threat to arriving personnel.
4. Assume command, implement the incident management system, establish a command post, and designate the appropriate sectors and officers based on the scale of the incident.
5. Summon the appropriate agencies, including the dive team, EMS, and the police.
6. Secure and manage the scene, ensuring that no unauthorized person gets onto the site. He should demarcate the cold, warm, and hot zones, as well as the staging area. This latter task includes ensuring free access routes for the various response vehicles. If necessary, he should plan for a helicopter landing zone.
7. If higher-trained personnel have not yet arrived, begin interviewing any witnesses. Otherwise, he should turn any witnesses over to law enforcement personnel.
8. Draw a profile map of the area.

At the Awareness level of training, personnel may establish the hot, warm, and cold zones.

Since diving incidents aren't a regular occurrence, it's recommended that Awareness-level personnel keep a checklist of their basic responsibilities in their car or wallet. It may also be necessary to keep the map of possible staging areas in their vehicle, depending on the size of their jurisdiction.

Operational Level of Training

For water-related emergencies, Operational-level efforts consist of shore- or boat-based tasks. It is these members, the tenders and profilers, who essentially direct the course of a diving operation.

On arrival at the scene, Operational-level personnel should do the following.

1. Don the necessary personal protective equipment.
2. Conduct their own risk-to-benefit analysis, and inform the safety officer of any concerns.
3. Finish implementing the incident management system.
4. Create a profile map and determine the location of the victims, if not already done.
5. Interview witnesses and create a profile map, if not already done.
6. Identify available resources and ensure an adequate response.
7. Develop a shore- or boat-based plan of response based on the available resources, as well as the team's procedures and guidelines.

8. Assist the divers with their equipment and deployment into the water.
9. Serve as dive tenders and profilers.
10. Be constantly aware of any incipient dangers.

Tender: The tenders are the most important people on the site, for they have the greatest overall responsibility. It is the tender's job to take care of the needs of the diver. As such, he functions as the link between upper management and those in the water. A tender knows what's going on at a search scene. By the trail of bubbles, he can see where his diver is and where he's going; he knows the angle of the tether, the

By watching the trail of bubbles and minding the tether, a tender can monitor the status of the diver.

movement of the line, and how much line is out. A good tender will know within two to three seconds when the diver's tether line has snagged, often before the diver knows it himself. The same holds true for breathing rates and speed of movement. Thus, the diver has little to do or think about other than searching, keeping the tether line taut, and his own safety. The responsibility for all of his other concerns falls on the tender.

On arrival at a site, a tender should do the following.

1. Don a PFD and other personal protective equipment.
2. Take air cylinders, tending lines, and all other necessary equipment out of the rescue vehicle. As he lays them out in the staging area, he should turn the air on in all of the tank assemblies, check the pressures, and orient the full tanks so that their valves are facing the water.
3. Assist the diver into his exposure suit and gear.
4. Check the gear.
5. Maintain an organized, well-staged site. He should assist the diver while he is standing by, whether to provide drinking water for him, review the objectives, monitor him for stress, or otherwise.
6. Accompany the diver to the point of entry.
7. Receive a briefing from the dive coordinator before the diver enters the water.
8. Review line signals, safety procedures, and contingency plans before the diver enters the water.
9. Ensure that the backup and ninety-percent-ready divers are in place before allowing the diver to descend.
10. Ensure the safe entry of the diver, noting his time in and his starting tank pressure.
11. Direct the diver where to search, monitor the tether line, give signals as appropriate, and alert the profiler of any slack in the line.
12. Decide whether an area can be secured as having been properly searched.
13. Monitor the diver's air bubbles, record his breathing rate every five minutes, and continuously assess his status.
14. Assist the diver out of the water and through the dress-down phase.
15. Participate as appropriate in postincident debriefings.

A tender's attention must always be on his diver. This means he must never take his eyes away from the spot in the water where the diver is, even for a few seconds. Moreover, no diver should ever be allowed to enter the water without a designated tender, even if safety lines aren't being used. The rapidity with which a situation can go awry during water-related operations means that there must be no shortfall in accountability, and tenders must maintain their vigilance on a moment-to-moment basis.

Profiler: The task of the profiler is to record the progress of the search by keeping a map showing exactly where the divers have been and what areas have been covered. This map is one of the most important tools in public safety diving. Without the profile on paper, there is no way for surface personnel to know whether a diver has missed an area or not.

The job of profiler should be held by the backup tender; that is, the tender assigned to the backup diver. The reason for this is that profiling duty serves to heighten a backup tender's awareness of the dive operation. Since the profiler must record the diver's progress, his eyes will always be following the diver. He may even realize that there is a problem as soon as the primary tender does. Such a high level of involvement will allow him to deploy the backup diver that much faster and to the right location. It also means that a backup tender can readily be rotated into the position of primary tender, decreasing downtime and miscommunication. The backup tender can also make the best judgment as to whether the primary diver has performed an effective search or not and whether the search area can be secured. Another benefit of using the backup tender to act as the profiler is that it will allow the team to operate with fewer personnel.

Technician Level of Training

Only properly trained and certified divers belong in the water performing public safety diving. A sport diving certification doesn't meet the need. The Technician level of training requires extensive knowledge, skills, training, and special equipment. At an incident, diving technicians may be assigned one of three roles: primary diver, backup diver, and ninety-percent-ready diver.

Primary Diver: The diver's role in a low- or no-visibility environment is simply to search the bottom as directed by the tender. The diver is the tender's eyes. His purpose is simply to search at the end of a rope that he keeps taut, going wherever the tender directs him to go. Because this is tethered, tender-directed, solo diving, the diver has no buddy to distract him from the search.

As a team member, the primary diver's responsibilities include maintaining equipment, records, good health, good fitness, and diving skills. During an incident, divers must strive to reduce the risk to which they and other team members are exposed. On arrival at the scene, a primary diver should do the following.

1. Say no to any dive he feels incapable of performing.
2. Dress quickly with a tender's assistance.
3. Keep his conversations brief and related to the incident.
4. Conduct a proper gear check with the tender, and again later at the demarcation line with the safety officer, before entering the water.

5. Review the object, search pattern, line signals, and contingency plans with the tender.
6. Perform the mission as safely as possible while executing an effective search pattern.
7. Stay well-hydrated and maintain a proper body temperature during the operation.

If the dive is called off or discontinued for any reason, the diver shouldn't question the dive coordinator's decision at the scene. Such discussions are more appropriately held out of the public view, preferably at the team's headquarters.

A diver's responsibilities don't end once he exits the water. He must immediately report any problems or symptoms, whether of decompression sickness, air embolism, or otherwise, to the dive coordinator and EMS personnel. Even if he feels fine, he must be checked out by EMS personnel after dressing down. The diver is responsible for filing out a personal logbook and appropriate department paperwork. Equipment checks and maintenance are also within his domain. The primary diver is a key figure at the postincident debriefing.

Backup Diver: In addition to having the normal duties of divers, a backup diver must be ready to act as a replacement if the primary is unable to perform for any mission, and he must be ready to render assistance if the primary runs into trouble. The backup is also the next in line in the normal rotation of divers. For these reasons, a backup diver should monitor the progress of the search as reported by the tenders. He should be constantly alert to any circumstance that might demand his participation.

To be fully ready to meet these responsibilities, a backup diver at the scene of an emergency should do the following.

1. Have all of his gear, except for the primary regulator, in place.
2. Have the contingency line in place and ready for attachment.
3. Be properly weighted to prevent any problems with descents.
4. Be mentally prepared to deploy at any moment in an expedient, disciplined, professional manner.
5. Be fully capable of performing well-practiced and tested contingency plans, using practiced communication techniques.

Ninety-Percent-Ready Diver: Because of the complex nature of diving, it's always possible that the backup diver will experience a problem when called. Likely problems could range from equipment failure to inadequate weighting to the loss of the weight belt to an inability to equalize. Following a policy of having contingency plans in place, it's best to have a second backup diver available, wearing an exposure suit and with his gear fully checked and functioning. If the backup diver is called on to make the descent, the ninety-percent-ready diver completes the dressing process so that he is fully ready to enter the water. It is often easier and faster to deploy a ninety-percent-ready diver than it is to reweight or make equipment changes on a backup diver. With a ninety-percent-ready diver in place, the redundancy and safety of an operation increase dramatically.

Because a ninety-percent-ready diver replaces the backup, he doesn't need his own tender if there aren't enough personnel present or if the dive platform is too small to accommodate more. The backup tender operates with whichever diver makes the descent.

Selecting Personnel

Skill levels vary considerably among divers on the same team, typically because of different experiences, certification, training, practice, or even personal comfort in the water. These discrepancies in skill often mean that the same people are chosen for the same job every time. The stronger, more comfortable divers are often chosen as the primaries, whereas those who are less at ease in the water are often held in reserve as the backups. If your team follows that line of reasoning, you may wish to consider the inevitable question it poses: If a stronger diver cannot get himself out of a predicament underwater, how can you expect a less-skilled diver to enter the same environment to help him?

A strong, skilled diver will go the farthest into the weeds, a submerged vehicle, or debris. If such a diver gets into trouble, chances are high that a less-skilled diver won't even be able to reach him, much less bring him out. Too often we believe that nothing will happen to the primary, but reality demonstrates otherwise. Entanglement, equipment problems, and air depletion do occur, thus necessitating backup divers who can handle themselves well underwater.

Any emergency is a stressful situation, and the stress level can increase exponentially for a backup diver who may be less than comfortable wearing scuba gear or even in venturing below the surface. Since stress leads to error, an unconfident member will immediately be more prone to such problems as hyperventilation, indexterity, an inability to equalize, mental confusion, or simply succumbing to fear. Using a stronger diver as the backup, however, means that, should any of these problems occur for the weaker diver, the stronger one will be fully prepared to help. By this reasoning, the ninety-percent-ready diver should be the strongest of all, for he is the one designated to make that last-ditch effort to save the primary's life.

A primary diver should only go as far as his backups can go to get him out. If you don't fully believe that your backup diver can save you, then you need another backup. If you, as a backup diver, don't feel capable of rescuing the primary, then you aren't ready to be a backup. Be honest in these evaluations, for they could mean your own life, as well as that of another diver.

One final word on this subject. Rotating your weaker divers into the frontline position will give them the experience and, ultimately, the confidence they need. The overall effect of this practice will be that the entire team becomes stronger. Naturally, you should never put weak divers into positions that are beyond their personal capabilities.

After reading this chapter, you may feel as if dozens of people are needed to run a safe diving operation, but this isn't the case. It's obvious that at least one diver is necessary, and since most public safety diving is directed by the tenders, then a tender is also required. A diver cannot be sent down without another one standing by, so a backup is mandatory, as is his tender. Finally, to step in when unexpected problems befall the backup, a ninety-percent-ready diver should also be considered essential. This means that a minimum of five well-trained, certified, properly equipped personnel—three divers and two tenders—are required to conduct a single-dive operation.

If the weather and water conditions cooperate, then the five-member response team described above is also adequate to support a second dive. In this scenario, the backup tender and diver become the next primary team. The ninety-percent-ready diver becomes the backup diver, and the previous primary tender becomes the backup tender. The previous primary diver, after rehydrating and being quickly checked out by EMS, changes tanks and becomes the ninety-percent-ready diver. As long as conditions permit, and as long as these personnel don't become fatigued or too cold, then a third, fourth, fifth, and sixth dive can be made by continuing this rotation process. In the rotation, divers should have at least forty-five-minute surface intervals. That surface interval will conveniently occur as the other two divers make their twenty- to twenty-five-minute descents. The divers must be sure to stay within their no-decompression profiles.

When only five people are in operation, the senior tender can also function as both the incident commander and the dive coordinator. To ensure that this minimum contingent responds to any given emergency, a team should probably be comprised of at least nine members, or six divers and three tenders. That way, if only about half the team shows up, it can still be operational.

This number of five as a minimum contingent also applies to high-visibility operations. For example, when using a tow bar, a device that is pulled across the search area by a boat, you need two divers on the bar, each acting as the other's backup, plus another backup diver on the boat, a tender, and the boat operator.

All divers should be trained as tenders, but not all tenders need to be divers. Nondiving tenders gain more tending experience and often take the role more seriously. Department members with many years on the job, but who may be too old to act as public safety divers, can make excellent tenders, particularly since they bring their experience to the team. EMS personnel can also make particularly effective tenders, since they are accustomed to observing patients and continually recording information. They can also help shave time off the operation by performing the diver checks themselves. Let it be said that a team that requires all of its members to be divers will not only waste money, it will also exclude some personnel who would otherwise make excellent tenders.

A safe diving operation doesn't need to be a big operation, and an effective team doesn't need to have scores of members. The right combination of a dedicated few will prove far more valuable in the long run.

Study Questions

1. What two general aspects of teamwork are required of dive team members?

2. True or false: Some state statutes require that the moment a police officer is on the scene, he must take control, whether or not he is trained in the needs of that operation.

3. To ensure fitness for a team, the goal of a public safety diver is to be able to swim how far in ten minutes without any swim aids?

4. To ensure fitness for a team, tenders and support personnel should be able to swim how far while wearing a zipped PFD?

5. Certification to the Awareness level usually involves at least how many hours of training?

6. Are Awareness-level personnel certified and expected to interview witnesses if no other personnel are present at the incident site?

7. Which team members have the greatest overall responsibility at a dive site?

8. When should a tender record the breathing rate of his diver?

9. What is the basic task of a profiler?

10. The job of profiler should be assigned to what member?

11. What is the primary diver's role?

12. What is the advantage of using a tethered, tender-directed, solo method of search?

13. Why should a backup diver monitor the progress of the search as reported by the tenders?

14. Why doesn't the ninety-percent-ready diver need his own tender?

15. Why shouldn't the strongest members on the team act as the primary divers?

16. What is the minimum contingent of personnel that can safely mount a single-dive operation?

Chapter 4

Diver Training

What does it take to become a public safety diver? It takes the right preparation and equipment, of course, but even more is necessary. It all begins with proper training. Too often, candidates for public safety dive teams receive entry-level recreational scuba training, put on a patch, and go into operation. Yet no fire department would allow someone with a conventional driver's license to drive apparatus, and no EMS squad would allow just anyone to drive its ambulance. That the basic requirements for public safety diving should be as inexact as they are defies all logic, particularly given the dangers inherent to underwater operations and the culture of safety for which these teams exist. An open-water scuba certification is just a beginning. It doesn't prepare anyone for the exigencies of the rescue/recovery mission.

Why, then, are so many dive teams allowed to become operational with only recreational training? The availability of training is one reason. In firefighting, one cannot simply go down to the local haz mat shop to get training for hazardous materials, and you don't learn high-rise firefighting from the local high-rise firefighting store. Many, however, believe that they can go down to the local recreational scuba store to purchase training and equipment for public safety dive team personnel. Typically, the leading members will go down to the local dive store and begin bargaining. "If we train six divers and buy six sets of gear, what kind of discount can we get? Can we get free training?" Because of limited budgets, the team buys what the store will sell for the least amount of money, even though the salesperson probably has little or no understanding of the needs of public safety divers. Similarly, the training is conducted by the shop's sport diving instructor, who likely has no experience whatsoever with emergency response. Also, because the salesperson lowered the training price, the store can't afford to offer more than what would normally be given for a sport training course. The team members come away with an initiation to the realm of recreational diving, which is no more than what they paid for.

The Essentials of Proper Training

Step One: Basic Open-Water Scuba Certification

The first training you'll need on your journey toward becoming a public safety diver is an entry-level open-water scuba certification course. Such a course will teach you the basics of scuba diving and allow you to rent or purchase scuba equipment. The training you receive will depend on the certifying agency and instructor; however, the initial certification should complete all of the training requirements for open-water diver for any one of the recognized training agencies, including American-Canadian Underwater Certifications (ACUC), the National Association of Underwater Instructors (NAUI), YMCA Divers, the Professional Association of Diving Instructors (PADI), Scuba Schools International (SSI), and the National Association of Scuba Diving Schools (NASDS). Some certifying agencies do not allow instructors academic freedom to alter their entry-level curriculum, a policy that has both pros and cons. An advantage of a static syllabus is that each and every course is targeted to meet a minimum set of standards. By comparison, courses taken with instructors who have more syllabus freedom can be either above or below the minimum standards of the more rigid courses.

The recreational course should involve at least fifteen hours of actual in-water pool time for a class of six students. Of those fifteen hours, at least thirteen should be underwater. You should only count the time spent performing skills in the water, not time spent on the pool deck or time spent talking on the surface. There should also be at least twelve to fifteen hours of classroom time. If the greater portion of the academic time is spent watching videos, you may want to think twice about enrolling in that course.

The following list represents skills in which students need to be particularly strong if public safety diving is their goal. All of these skills should be developed and practiced under the supervision of a certified, insured diving instructor.

1. Breathing and working underwater with a flooded mask or no mask at all. Public safety divers stand an increased risk of losing or flooding a mask as they search the bottom. Thus, future public safety divers need to be comfortable underwater, with or without a mask in place. They not only need to be able to clear it, they must also be able to perform other skills with the mask off or flooded. Also, if the divers will be wearing full-face masks, they must be comfortable breathing from a pony regulator without a mask in case they have to remove it to switch to a redundant air source. Although divers should be carrying a standard, half-face mask, it will take time to don that mask after switching to a pony regulator.

2. Performing proper ascents and descents without visual aids. Future public safety divers must be able to perform thirty-foot-per-minute free ascents and slow descents with their eyes closed, since they will be operating in black water. Prospective divers should first practice ascending and descending with a buoyed, weighted line with a knot tied every foot. When they can repeatedly make ascents and descents at the proper rate, they should then practice with their eyes closed. The next step is to practice with straight, unknotted line, and then with no line at all. Finally, they should make ascents and descents while holding a ten-pound weight to simulate a lifeless adult body, which weighs approximately nine to sixteen pounds underwater. Be sure that the weight you use won't easily slip from a diver's grasp, which could result in a too-rapid ascent. Also, as will be discussed later, a diver should never use his buoyancy control device as an elevator while making an ascent.

3. Moving without sculling and kneeling without moving. Neither sport nor public safety divers should be allowed to scull with their hands, but this is especially true of public safety divers, who need to use their hands for search and recovery. Sculling and arm waving also increase air consumption and heart rates. If divers are allowed to scull during training, they will continue to do so after certification, which will make them less effective as searchers and less safe divers in general.

When divers kneel on the bottom, they must be able to remain still without any hand or arm movement. Public safety divers work on the bottom. If they become entangled, they need to be able to stop, kneel, and work with both hands without falling over. They need to perform other tasks on the bottom as well, and eventually need to be able to pick up and carry evidence containers or the body of a victim.

4. Being familiar with the equipment. Divers must be able to reach and use every piece of equipment on them, and they must be able to do so by reflex, without looking. Since the majority of public safety divers work in areas of low visibility, that need is critical. During training, they must learn to fix and manipulate their equipment while wearing a blacked-out mask. Moreover, since the life of a

During training, divers must learn to fix and manipulate their equipment while wearing a blacked-out mask.

public safety diver may depend on the adroitness of his backup, divers must all be trained to make equipment adjustments on each other as well. Examples of such tasks include being able to adjust a fin strap, hook up a pony regulator, and refasten a weight belt. Every student in the class should have the same equipment configuration, and it should be the same configuration that they will eventually use during operations.

Being able to replace a lost regulator is a life-and-death skill. Only the over-the-shoulder retrieval procedure should be used, both at the surface and underwater. The more commonly taught side-sweep method only works most of the time. Should the regulator be caught behind the diver, the side-sweep method will fail, and the chance that a diver will reflexively go to the over-the-shoulder procedure, which was only performed a few times in class, is very unlikely. The more likely result will be a bolt to the surface. The over-the-shoulder method also offers the lifesaving benefit of making a diver capable of turning on his own air in the water, in the event that he somehow skipped his equipment checks before entering.

5. Being able to manage entanglements. Entanglement is the greatest hazard that public safety divers face, because it is the only one that can prevent them from returning to the surface at will. Thus, training to deal with this problem is crucial from the start, where it can be introduced in a safe, controlled environment. Divers should

Entanglement is the greatest hazard faced by public safety divers.

build a cutting station from a weighted milk crate, plus fishing line and other snare hazards, to be placed on the bottom of the pool. The instructor can then run drills in the art of cutting. From these drills, the divers will learn how to cut different types of entanglements from the crate, first with their eyes open, then with their eyes closed. They will also learn that shears and wire cutters are the tools of choice, rather than knives.

6. Checking gear and making their equipment trim to the body. Tenders should be involved in the initial training so they can be taught how to outfit a diver properly, ensuring that the equipment lies close to the diver's body, thereby reducing the chances of entanglement.

7. Ditching weights and gear. Weight ditching on the surface is an important skill for any diver, but the importance is paramount for public safety divers. For example, a diver may find himself in a current on the surface and need to ditch his weight belt without losing hold of a recovered body. Students should practice ditching weight belts while wearing full gear at least five times during every confined-water session. If the instructor is concerned about damaging the bottom of the pool, the students may practice ditching empty weight belts. Because of limitless entanglement scenarios, public safety divers must also be able to ditch and don their equipment underwater in conditions of zero visibility, so students should practice these skills using a blacked-out mask.

8. Developing proper kicking technique. All divers need a good, strong kick to be able to cope with varying currents and for a host of other reasons. Public safety divers need an especially good kick, since they must sometimes skim along the bottom at a brisk pace for twenty minutes or more during a search. Anyone who might recover a body must be able to propel himself efficiently. An inefficient kick can lead to severe leg cramps that can bring an entire rescue operation to a halt and result in fatigue that puts the diver at risk. The majority of entry-level, recreational dive courses don't produce divers with kicks that are effective enough for a public safety mission. Probably more than half of the nonprofessional divers out there suffer from some form of shallow, bent-knee, limp-ankle kick. These types of deficiencies are bad for sport divers and worse for public safety divers. Instructors for any open-water course should have their students learn and practice a full, effective, straight-legged, pointed-toe kick from the thigh. Once the students can perform a strong flutter kick, they need to learn how to perform kicks that won't stir up the bottom if the water has a foot or more of visibility.

9. Exiting the water. Most public safety dive teams work off of small boats, such a fourteen-foot inflatables, which have no ladders. Thus, divers need to train well for deep-water exits. Such an exit is accomplished by first removing the weight belt, then the buoyancy compensator device. The diver then kicks his way out of the water. For other exits, the weight belt should be ditched and passed off to a tender before any other piece of gear is removed, whether exiting into a boat or onto the shore. The mask and regulator must be kept in place at all times until the diver is completely out. Because of contamination hazards and for other reasons, public safety divers need to practice such procedures until they become habit.

After divers receive their open-water certification, they should only be considered to be apprentice-level divers. As apprentices, they may work with the dive team, but under no circumstances should they be allowed to dive during actual operations, since they aren't yet trained as public safety divers. Note that unless apprentices have received training as tenders, they shouldn't be allowed to serve as tenders, either.

The mask and regulator must be kept in place until the diver is completely out of the water.

Step Two: Becoming a Public Safety Diver

Once divers have their open-water certification, they should dive as much as possible, and even sport diving will give them time to become increasingly comfortable and capable in the water. For public safety diving, however, practice alone isn't enough. New candidates should first pass a public safety rescue/recovery course before they can act as public safety divers.

Every diver should take a sport rescue course at one time or another, because the skills they'll acquire in such training are important. However, a sport rescue course won't cover the diver-to-diver rescue skills necessary for public safety diving, and it won't be relevant to public safety search operations. The emergency procedures in a sport diver course are based on buddy techniques while diving at a midwater level in conditions of good visibility. If an unencumbered public safety diver uses up his main cylinder, he can switch to his pony bottom and ascend to the surface. Unfortunately, the other scenario is one in which he is entangled or trapped in a confined space on the bottom. None of these issues are addressed in sport rescue programs. The diver-to-diver skills taught in such courses will, however, be useful for dive teams that operate in clear water and perform searches with divers in tandem.

Most low-visibility public safety water operations require tender-directed, tethered, solo diving. This technique is safer, faster, more accurate, and more effective

Although an important facet of the training process, a sport rescue course won't cover the skills necessary for public safety diving.

overall than techniques involving multidiver lines or untethered search patterns. Sport rescue and recreational search and recovery courses teach the latter two methods. The techniques taught in most of these programs are less effective for low-visibility searches for an adult body, and they don't even begin to address the procedures required for small objects, which public safety teams might be called on to find.

If a public safety team has a limited budget and has to decide on which course to take, the choice should be a good public safety diver rescue/recovery course that includes plenty of diver rescue skills, hands-on contingency plan training, and accident prevention procedures. A team can move to operational status with such a course, but it can't do the same with only sport diver rescue training.

Once divers pass a public safety diving course, they should be considered to be Diving Technicians, Level I. Teams may wish to create a probational status for new Technician-level divers, under which they may dive and serve as backups during drills but not actual operations.

Step Three: Continuing Education

A Level I public safety diving course doesn't mean the end of training for a true public safety diver. In fact, it is only the introduction to a career, whether paid or volunteer. After the original course, public safety divers must maintain their competence either by recertification every two years or by taking continuing education courses in the field. For example, if the divers weren't originally trained in dry suits, a dry-suit training course would be a good next step for continuing education. Beyond that, there are a multitude of public safety diving specialties, including large-area search, underwater vehicle extrication, ice diving, and others.

Drills

Public safety divers should avoid on-the-job training. You don't want to be performing new techniques or trying out new equipment at the scene of an actual emergency. Drills and team practice are an essential part of continuing education. Ideally, drills should take place monthly in the form of a mock rescue or recovery operation in areas in which your team is likely to be called to dive. Monthly practice, of course, isn't always realistic, especially for municipal departments that must then budget for overtime. In such cases, practice as much as the budget will allow. Also, don't allow recreational dives to count as drills. Any in-water time can be valuable, but there is no substitute for practicing public safety diving techniques and contingency plans.

In the realm of sport diving, the standard is to take a brief refresher course if you haven't made any dives for six months. Because of the extreme nature of public safety responses, divers who haven't gone below in three months shouldn't be allowed to operate in the water on an actual call until they have first participated in a drill session.

Instructors

Who should train divers? In many areas, dive team personnel are trained solely by in-house trainers who are members of the team. An in-house trainer may be a senior teammate who teaches his fellows during drills, or he may be a certified scuba instructor who teaches members using a nationally recognized training program. Such training can have some benefits, and having a certified, insured instructor on the team is certainly a good idea; however, the sole use of in-house trainers isn't without drawbacks.

Although some teams trained this way have managed quite well, others have faltered and failed. Often, in-house instructors are simply recreational scuba instructors who happen to volunteer or are asked to teach the dive team. Hence, by default, those recreational instructors become public safety diving instructors, even if they have neither formal instructor training nor certification in the discipline. For this reason and others, in-house training is often substandard. The students may never receive written and practical exams, and they may never be required to complete a minimum number of classroom and in-water hours. All of this means that in-house training provides less legal accountability, raising a serious liability issue. In any lawsuit against the team, lawyers and experts will come out of the woodwork, questioning whether the team operated according to standards set by professional public safety diving certifying agencies. A team trained solely in-house may be hard-pressed to demonstrate that its standards were suitable.

Like all aspects of emergency services, public safety diving is constantly advancing. Without outside trainers to introduce state-of-the-art techniques and equipment, a team may stagnate. Simply put, in-house trainers who only train local teams don't gain the experience, knowledge, and skills that professional instructors acquire by working across a wider spectrum. Rather, unsafe, bad, or sloppy procedures may become the company standard. Also, because they're too familiar with their trainer, the members of a given team will tend not to take their in-house trainer as seriously as they would an outside instructor. As a consequence, the trainer, students, and the team as a whole may never reach their full potential.

In-house training does have some merits, particularly in terms of cost and ready availability. Because of the many associated pitfalls, however, it's best when it's supplemented with periodic training from outside professionals who can introduce the latest innovations and objectively ensure that the team is operating as safely as possible. In-house trainers should take outside training courses at least once every two years to update their knowledge and maintain their proficiency.

The Limits of Specialty Training

At various junctures in this text, it has been mentioned how sport diving experience and certifications are no substitute for true public safety diver training. The converse of that statement also holds true: Personnel trained as public safety divers may not be fully prepared for sport diving or even emergency responses outside of their particular expertise. In fact, numerous public safety divers have died not on the job, but while attempting sport dives for which they had no training or experience.

Most public safety divers operate in conjunction with a surface team, replete with backups and contingency plans. Hopefully, the dive plan itself has been created from a set of well-practiced procedures and is overseen by an experienced dive coordinator. A diver's gear is thoroughly checked by his tender. When the diver makes his descent, it is typically into shallow water of thirty feet or less. The diver is accustomed to being tethered. His tender tells him where to dive, when to descend, and when to resurface. He is used to working on the bottom, not hovering in midwater. Generally he's overweighted by two or three pounds.

In contrast, sport divers have no tender or backups to rely on. There's no tether to tell them where to go and to connect them to a surface team. Sport dives are often done in a location far from medical help, and there's never an EMS crew standing by. Any dive planning is up to the buddy team. Sport divers monitor their own air, depth, and time, and they have plenty of practice with midwater buoyancy control.

The lesson contained in these distinctions is simple. Just because you're a great black-water diver doesn't mean that you can go out and sport dive safely. Public safety divers should limit themselves to environments and the types of dives for which they are trained.

The diving industry has set parameters as to how deep different levels of divers may go. Novice sport divers are trained on the premise that they are capable of safely going to a maximum depth of sixty feet. An open-water certification doesn't prepare you for descents in excess of that. Even with proper training, a diver should have logged at least twenty-five dives before attempting deeper descents.

A standard public safety diving course will train you to operate within a given depth range, typically to a maximum of fifty feet, with a possible extension to sixty if permitted by the dive coordinator and the safety officer. Research shows that the average public safety team in the United States is made up of open-water divers with little or no further dive education. Some teams have senior members with fewer than fifty total lifetime dives. Entry level is entry level, no matter whose course you've taken. If it is your team's intention to be able to conduct deep-water operations, you must begin with a solid base of experience, then seek the appropriate training. Work to depth gradually.

Training topics that need to be strongly reinforced and updated for deep operations include planning and monitoring air consumption; carrying larger-capacity bottles; planning no-decompression profiles; free-descent and free-ascent skills; hovering; contingency decompression procedures; proper weighting; and thermal protection. A team should also consider training with surface-supply systems.

The issue of commercial diving is equally problematic for public safety divers. Several have died attempting to perform jobs that should have been done by trained professionals. In the hopes of saving money, companies will occasionally approach a dive team to perform an underwater job of one kind or another. Happy to have the work and usually in need of funding, teams sometimes accept these offers. Because the members may not realize the complexities of submerged vehicle operations, dam inspections, and the like, many won't even be aware of the dangers they might be facing.

It's true that the Occupational Safety and Health Administration doesn't regulate diving that is "performed solely for search, rescue, or related public safety purposes by or under the control of a governmental agency." However, any side jobs that a public safety diving team takes on may technically be required to meet OSHA standards. A team that doesn't meet those standards faces the possibility of massive fines from the agency.

Commercial training usually requires a minimum of one year's experience, full time, plus one or more years apprenticing as a tender. If your team is asked to perform a job that should otherwise go to a commercial diver, you should defer based on an honest assessment of your capabilities and limitations.

Training Exercises

Dry-Land Searches: Begin your search training on dry land. Have your divers don their harness and buoyancy control device. Cover their masks with duct tape so that they cannot see. Attach a tether line, and hide a small object sixty to eight feet away from the tender's position. The tender should let out the diver to the search area, but he and the diver must not speak to each other—all communication must be done using line-pull signals. In one variation of the exercise, the tender is also unaware of the object's location. It's beneficial to deploy several teams simultaneously, since the tenders will then have to communicate with each other and be aware of the divers in the adjacent search areas. The goal for completion of this operation is ten to fifteen minutes, and up to twenty if the tender is also being trained.

Full-Face Mask Training: The first procedure when using any mask is to acclimate your face to the chill of the water before you go in. Since a full-face mask will

The complexities of a full-face mask warrant special training.

keep your face warm, it will leave you more prone to laryngospasm if it should suddenly become dislodged. Moreover, since a full-face mask has an integral regulator, the skills needed to don, remove, replace, and clean it are different from those required for a half-face mask. These differences mandate special training.

All full-face masks have a strap system, called a spider, with five or six straps connected together at the back of the head and attached to the mask at the top, sides, and bottom. To don the mask, loosen the straps and pull the spider in front. Next, place the mask on your face. Making sure that it has a good seal, go ahead and begin breathing from it. Holding the mask with one hand, flip the spider over your head with the other, then tighten the straps. Start with the straps on the sides, nearest your temples, then tighten the bottom straps, and finally the top ones.

The top straps should have no more than two inches of excess past the buckle. If there is more, it means that they are too tight and that the bottom straps haven't been secured properly. This will cause the chin of the mask to lift up too far when the diver looks up, causing the mask to flood. When all of the straps have been tightened properly, the spider should sit squarely on the back of the diver's head.

When removing the mask, grasp the chin portion, then lift up and away, sliding the mask over the back of your head. Although flipping the bottom buckles to loosen the straps before lifting up will make removal easier, it isn't necessary. Removing a mask in this manner will allow you to make a quick switch to your pony regulator in the event of an emergency. Whenever the mask is off your face, be sure that you are exhaling, since holding your breath can lead to lung injury. Also, when removing a positive-pressure mask underwater, be sure to flip the shutoff switch before you break the seal of the mask to your hood. Otherwise, the mask will free-flow violently, wasting air.

Switching to a Pony Regulator: Changing from the full-face mask to a pony regulator (not an octopus) is the most important skill for using a full-face mask. Because this skill helps prevent out-of-air emergencies, practice it until you can do it completely by reflex. Ice divers should note that switching to a pony regulator can also be critical in situations where a freeze-up will cause a full-face mask to free-flow. Many teams require that all divers carry a second, standard face mask, usually worn backward around the neck. You should practice switching to the backup mask as well when you practice switching to the pony.

Clearing a Full-Face Mask: A full-face mask should be cleared as soon as you put it on your face. Do not wait until you have set and tightened all of the straps; otherwise, you may find yourself short of breath. Clearing the mask is accomplished by pressing the purge button on the mask's regulator and holding it until the water has drained out. Depending on the type of mask, you may need to keep it in an upright position for it to drain properly. As mentioned above, a positive-pressure mask will

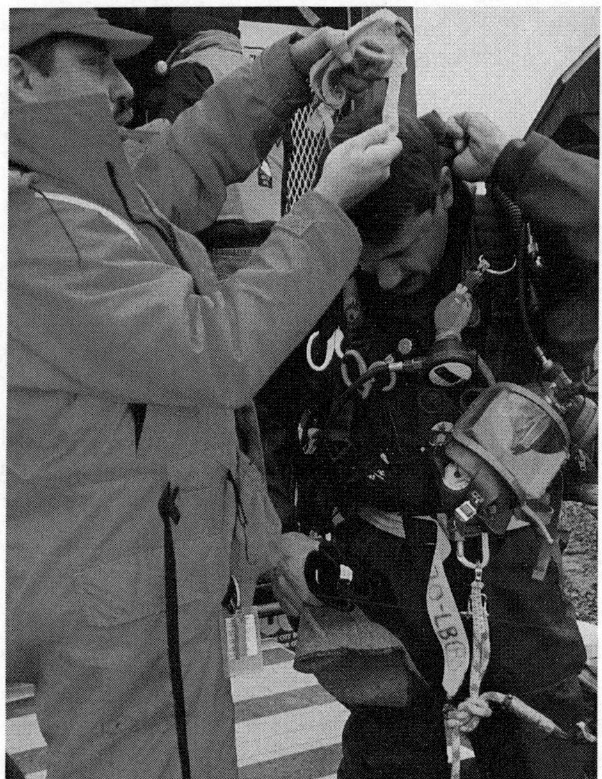

Many teams require that all divers carry a second, standard mask, usually worn backward around the neck.

free-flow violently when removed from the face unless the shutoff switch has been flipped. Because of the force of the air, it is difficult to reseat, seal, and clear such a mask unless the flow has first been curtailed. Once the mask has been sealed and the switch opened again, the flow of air will force the water out, thereby clearing the mask.

Many divers feel that there is no point in clearing and replacing a full-face mask, since they would be switching to a pony regulator and a half-face mask anyway in the event of a mishap. Still, the ability to replace and clear a full-face mask is most valuable. First, if the water is contaminated, the diver won't have to put a contaminated mouthpiece into his mouth. Second, simply replacing and clearing a full-face mask will allow a diver to continue with the dive rather than having to abort. Third, replacing and clearing a full-face mask means that the diver can continue breathing from his main tank, which is important if the pony bottle has already been donated to a diver in distress.

All of these skills must be practiced while wearing gloves. It should be mentioned that many teams use transfer blocks, which allow a diver to switch between air sources while using the same second-stage regulator and full-face mask, thereby avoiding exposure to frigid water or contamination. Divers still need to practice with these devices in a controlled environment, however. They also need to practice all of the other skills, since even a transfer block is no guarantee that a mask won't become dislodged or that a diver won't have to rely on a contingency tank.

Entanglements: Dealing with entanglements should be part of your regular training drills. Tie fishing line, wire, cable ties, spider wire, and any other likely snare to weighted milk crates. Have the divers cut pieces of each type underwater while wearing gloves. Also have them tie different types of knots to connect the pieces. Add further realism by blacking out the divers' masks. Cutting these hazards without being able to see them will definitely make them appreciate the value of shears.

Study Questions

1. What is the first step toward becoming a public safety diver?

2. Name at least two of the recognized training agencies mentioned in the text.

3. An entry-level course should involve a minimum of how many hours of in-water pool time, and how many of those hours should be underwater?

4. How should future public safety divers practice making proper ascents and descents without visual aids?

5. According to the authors, what is the greatest hazard that public safety divers face?

6. How should public safety divers practice ditching and donning equipment underwater, simulating conditions of zero visibility?

7. Once divers pass a public safety diving course, they should be considered to be ______________.

8. Generally, why is it preferable to use outside trainers as opposed to in-house trainers?

9. Novice sport divers are trained on the premise that they are capable of safely going to a maximum depth of ________.

10. True or false: Any side jobs that a public safety diving team takes on are exempt from OSHA standards.

CHAPTER 5

Equipment for Shore Personnel

When the members of a dive team begin considering the equipment they need, usually scuba gear is the first thing that comes to mind. Yet divers aren't the only ones with jobs to do or safety to consider. Shore personnel also require a fair amount of equipment. The following items are standard for tenders in modern operations.

PERSONAL FLOTATION DEVICES

The most essential piece of gear for shore personnel is the personal flotation device. No one on the emergency response team should be allowed within twenty-five feet of the waterline without one. If a slope leads to the water's edge, PFDs must be worn from the top of the slope, even if the top is more than twenty-five feet from the water's edge.

Although this might sound extreme, consider what might happen if an on-site accident were to render someone unconscious. Without a PFD, it's more likely that your head will submerge when you fall in the water. Without a PFD, a person could sink underwater in seconds, perhaps unnoticed. Even an expert swimmer can drown in twenty seconds if the water is cold, if a head or spinal injury is involved, or if he's wearing something nonbuoyant, such as turnout gear or a gun belt.

There are numerous styles of personal flotation devices, categorized by the United States Coast Guard into five different types. In Europe, the new CE standard recognizes four classes of PFDs, categorized as Buoyancy Aid, with 50 newtons of lift; Permanent Buoyancy Life Jacket, with 100 newtons of lift; Offshore Life Jacket, with 150 newtons of lift; and the Serious Life Jacket, with 275 newtons of lift.

Type I PFDs: (Offshore Life Jackets, with a minimum of 22 pounds of buoyancy for the adult size.) Type I PFDs are designed for offshore use, especially for rough conditions or when there's a chance of a delayed rescue. These bulky vests have the greatest flotation of all types and are designed to turn most unconscious persons to a face-up position. Type I PFDs are made in two sizes: child (under ninety pounds) and adult. Because of their bulk, these types are a poor choice for dive team personnel and boat operators, since they make it difficult to reach around a victim or to climb back into a boat if necessary.

Type II PFDs: (Near-Shore Buoyancy Vests, with a minimum of 15.5 pounds of buoyancy for the adult size.) These are similar to Type I PFDs, although they're designed for calm, near-shore conditions, where a rescue should occur quickly. They are less bulky than Type I PFDs and provide less flotation. A Type II vest will turn some persons to a face-up position. They are made in four sizes.

Type III PFDs: (Flotation Aids, with a minimum of 15.5 pounds of buoyancy for the adult size.) These are designed for continuous, comfortable use, thereby allowing the wearer to engage in activities such as swimming, water skiing, fishing, kayaking, and canoeing. Generally, they provide less flotation than a Type I or Type II PFD. Some won't turn the wearer to a face-up position, and some wearers may actually have to tilt their heads back to keep from going face-down. A Type III PFD may not have enough flotation to keep a wearer's face dry in rough waves. They are available in a variety of sizes and styles.

Type IV PFDs: (Throwable Devices, with a minimum of 16 to 18 pounds of buoyancy.) A Type IV PFD is designed to be dropped or thrown to a person already in the water. They are best when used in conjunction with wearable PFDs. Type IV PFDs may be rings, cushions, cans, or horseshoe floats. It's interesting to note that ring buoys were never meant to be thrown to a victim in the water. They were designed to be dropped in the water to someone who'd fallen off of a large vessel. The victim would swim to the ring buoy while the ship made a large turn around to retrieve him. If a ring buoy is thrown to a victim and it hits him in the head, it could be the last insult that makes him go down. Also, ring buoys don't deploy very far or accurately, especially if their lines are coiled. If you want to increase the effective range of deployment, use the ring buoy with a line-bag attachment. Line deploys faster, farther, and more accurately out of a bag than from a coil, as demonstrated by rescue rope throw bags.

Type V PFDs: (Specialty-Use Devices, with a minimum of 15.5 to 22 pounds of buoyancy for the adult size.) These PFDs are designed for specific purposes and may be used for those purposes. Examples include whitewater vests, deck suits, and hybrid (inflatable) PFDs. Flotation coats and full suits are excellent for public safety personnel working near water in cold weather, since they offer complete waterproofing protection, insulation, and flotation.

Of the five varieties, Type III and Type V PFDs are generally the preferred choices for public safety teams, since they provide both sufficient flotation and freedom of movement.

A few general words about flotation devices are in order. A PFD must be the right size for you, and if it isn't zipped or buckled properly, you aren't wearing it. An open PFD will quickly come off in the water. Many flotation devices have pockets, holders, and sizing straps, and these prove highly useful during operations, as do built-in harnesses with attachment points. Some tools that you should wear on a PFD include a water-activated flasher; shears; whistle; timer; note pad and pen; tie wraps; flashlight or penlight; eight to ten inches of duct tape; snorkel keeper, and a scuba tool for making minor repairs. When diving to make an extrication from a vehicle, the diver should carry a centerpunch for breaking the windows. Finally, avoid wearing turnout gear around water during incidents that don't involve firefighting. If the incident demands that you wear it, put on a PFD also—one that you have already tested in a pool with the bunker gear. Wear the PFD *under* the coat, since few of these devices can withstand the high temperatures of a fire without melting. This will also allow you to shed the coat if you happen to fall in the water.

Personal Protective Equipment

Tenders should wear head protection as appropriate to the weather, and they should wear a light, protective helmet when working around currents in excess of two knots. Any line handling should be done with gloves, which should also be appropriate to the weather. If the water is contaminated, wear rubberized gloves, if not latex gloves, under the work gloves. To prevent the work gloves from being contaminated, wear a second pair of latex gloves over the work gloves.

Your feet and legs should be protected from the sun, cold, splinters, glass, insects, and all else. Sandals and other open shoes are not acceptable. If you're working in an area where ticks may be a problem, wear high socks over your pant legs, and wrap duct tape around the tops. Rubber hip waders provide the best protection if the water is contaminated or marshy.

Insect repellent, sunscreen, and sunglasses are standard equipment for most dive sites. Drinking water should also be available for all personnel. Hands-free, backpack-type hydration systems, as used by bicyclists, are recommended.

Harnesses may be necessary if the tenders have to work on the ice, a slippery platform, a steep embankment, or anyplace with narrow footing. The harness should have a D-ring between the shoulder blades for tethering purposes. A good dive harness will do the job.

Other Equipment

Much of the nonapparel equipment needed at the site has to do with record keeping and management of the incident. If not a profile slate, then the backup tender will need at least a pad and pen to record information pertaining to the divers and the area searched. Those encharged with interviewing witnesses will also need writing equipment, as will the member logging personnel at the entrance to the hot zone. Clipboards and waterproof paper are helpful for all of these purposes. An incident command board is needed at the command post, showing information from the profile map, plus other vital data. Filled-out forms showing medical and next-of-kin contact information for each diver should be kept on the team's vehicle in case of an emergency.

Tarps are indispensable for the staging area, since they can provide a clean place for the divers to suit up, and they can also be used as protection against the elements, if necessary. You'll probably need police or fire tape to demarcate the warm and hot zones. To secure the diving area from boaters, be sure to use diver-down and alpha flags. If necessary, call in police boats. Portable radios demarcate the incident in another way. Not only do they facilitate communication between the members, particularly during large incidents, they also lend an air of professionalism to the management of the response. This may be important if crowd control or freelancing becomes an issue.

A shelter may be necessary, as dictated by the weather, and the quality of the water itself may mandate setting up a decontamination station. Emergency oxygen equipment and a first aid kit should be at the site, and ear-wash and mouthwash solutions should be available as well. During water operations, shore personnel should always be equipped with rescue rope throw bags, ready for fast deployment.

Lines

Because they are the sole connection to the world above, lines are among the most important pieces of equipment for public safety diving. Tether lines should be 3/8-inch in diameter, since that size offers the best compromise between strength and drag. Thicker diameters create more drag and a greater workload, and they increase the chance that the diver will turn too much toward the tender. Half-inch lines are sometimes specified in SOGs; however, that's usually because the recommendation was made by a high-angle expert rather than a public safety diver. High-tensile-strength ropes are too heavy, costly, and also unnecessary. Lines with 1,200 to 1,500 pounds of tensile strength are adequate for underwater rescue and recovery operations.

Some teams that operate in zero-visibility waters are currently opting to have their divers use small line reels, as are generally used by cave and wreck divers. Such a system has several problems. First, it puts control of the length of the line in the hands of the divers rather than the tenders, which decreases the ability to make a thorough search. Also, such lines are too thin to handle easily. Finally, it's difficult to attach such a reel to a diver's harness, which means that the diver could accidentally drop and lose the reel, thereby losing his direct-line pathway to shore.

Braided polypropylene is the best type of line for water operations. Because it floats and is soft to the touch, pliable, easy to knot, and easy to pack into bags, this type of rope is ideal for tethered diving. Also, braided polypropylene is dynamic; that is, it will stretch to a degree, meaning that a diver being pulled along the surface or into an obstacle will take less abuse than he would with static rope. Suitable 3/8-inch braided polypropylene will allow line-pull signals to transmit for more than 150 feet.

Laid polypropylene and nylon line are popular in the boating world, but they aren't good for tethered diving, since they aren't as flexible and hold knots poorly. Still, they do withstand wear and tear well, and they're relatively inexpensive. If laid rope is your choice due to cost or availability, it's better to splice the attachment loop for the diver rather than tying a knot. Also, put duct tape over the splice to give it greater holding power.

Although strong, kernmantle lines are expensive, and they tend not to float. As mentioned above, high-strength lines are unnecessary for tethered diving. Hemp and other natural-fiber lines are totally unacceptable for diving and water operations, since they are too absorbent and rot in them is inevitable.

Marking Line

Since a tender must always know where his diver is and what area has been searched, it's necessary to mark the lines for distance. Although there is no standard marking system, one widely used method, using narrow and wide wraps of duct tape, is shown in the accompanying diagram. Start with dry line and wrap the duct tape two or three times around. Using different-colored tape for each 25-foot interval will make the tender's job easier. Note that at 25, 50, and 75 feet, the small pieces of tape are on the tender's side of the wide pieces. Beyond the 100-foot mark, however, the

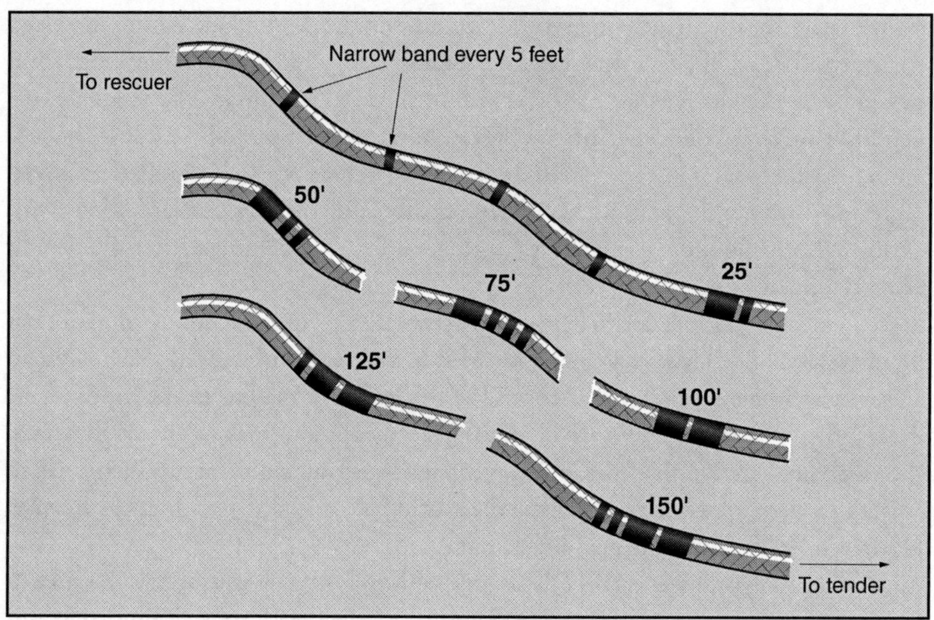

A rope-marking system.

small pieces are on the diver's side. This change serves as a visual reminder to the tender that the diver is more than one hundred feet from shore.

Besides tape, another option is to purchase prelabeled line, with numbers laced in by the manufacturer. Such lines are easy to use, but they can become wrongly calibrated if sections need to be trimmed or knotted. Also, they tend to be expensive. As an alternative, you can write numbers on the tape with a durable marker, or perhaps use a color-coded system for distance. In any event, lines should not be marked with knots or loops. Knots of any kind will weaken the overall strength of the line, and loops tend to become entangled with anything they find. Also, both knots and loops will seriously increase the drag on the line.

Line Maintenance

Checking for fatigue in the line requires both visual and tactile attention. Start at the looped end of the line and gently feel along its surface for weak, worn, or broken spots. The line should feel the same all over. Be alert for sections that feel thinner than others and places where the line feels as if it's separating in the center. Bad sections should be cut out if they're near the ends; otherwise, the line should only be used for nonsafety purposes. Don't tie a knot in the line thinking that you'll strengthen the weak section.

In braided line, look for feathers; i.e., sections where the strands appear as if they're sticking out. Check the inner core for weakness wherever you do find any.

Feathers alone aren't usually a problem, since they only indicate sections where the manufacturer spliced the line, leaving a little excess in the external coat.

You should never step on line, since that will weaken it. Boots and body weight will crush the fibers and grind in dirt, dust, and gravel, and ice cleats can sever it in one step. Always repack line promptly, even if it will soon be used again, to decrease the chances that someone will walk or drive over it. Keeping tether line neat, clean, and out from underfoot is everyone's job at an incident site.

To clean line, slowly dunk it up and down in a bucket of water, allowing the dirt to fall to the bottom. Another way is to use a line washer, which is a T-shaped assembly that connects to a garden hose. Run the line through the top of the T, allowing the hose to blow debris from the inner fibers. Otherwise, hang the line and rinse it slowly, since a stream at greater force will only wash away the surface dirt while further impacting the inner dirt.

To repack a line bag properly, hold the bag open with one hand and allow the line to pass through an O created by your thumb and forefinger.

Once clean, hang the line to dry. Don't try to dry line by laying it on the ground, since it'll only pick up more dirt and won't dry well. To dry properly, line needs to hang freely, with air circulating on all sides of it. Once the line is dry, repack it into a bag and store it in a cool, nonhumid place.

The use of rope bags has dominated the water rescue industry for years. Bags are easy to use and repack. They offer quick line deployment and help to keep the dive site neat and organized. Although reels can be used, they have a tendency to overrun, rolling faster than the rope can come off, causing the rope to become caught in itself. Reels with brakes are viable, but because they cost more and are more cumbersome to transport, line bags are still preferable. Coiling is the last choice for storing tether line, because it leaves the line unprotected and often promotes tangling.

To repack a line bag properly, hold the bag completely open with one hand, and allow the line to pass through an O created by your thumb and forefinger. With your free hand, pull the line into the mouth of the bag. When a bag is repacked correctly, it will release rope easily and snag-free every time. Additionally, if you wear gloves while repacking the bag, the gloves will clean dirt, sand, and other debris off the surface of the line.

Study Questions

1. No member of the response team should be allowed within ________ of the waterline without a personal flotation device.
2. The United States Coast Guard categorizes PFDs into how many different types?
3. How can you increase the effective range of deployment of a Type IV PFD?
4. Of the five varieties of PFDs, the two that are most preferred, for reasons of flotation and freedom of movement, are Types _______ and _______.
5. If you must wear turnout gear, where should you wear the PFD?
6. When should tenders wear a light, protective helmet?
7. Name two uses for tarps at a dive site, as mentioned in the text.
8. What diameter line offers the best compromise between strength and drag for tethers?
9. What is the best type of line for water operations?
10. Why shouldn't you use knots or loops to mark distances on a tether line?
11. What is the proper way to dry line?
12. What is the best way to store tether lines and why?

Chapter 6

Diving Equipment

All life originated in the sea, but water for humans is essentially an alien environment. That divers must bring along their own life-giving oxygen with them has few parallels in human endeavor, and there are fewer still in which the explorer must carry that air supply on his back. Aviators routinely ascend to altitudes in excess of two miles without the need for supplemental oxygen, but to venture anywhere below the surface is at once to go where oxygen starvation and drowning are ever-present possibilities. Whereas pilots measure their gains in thousands of feet, a mere sixty is already considered borderline deep for a free-swimming diver, and the pressure on his body at little more than half that depth is already twice what he experiences at normal ambient pressures on the surface at sea level. It's sad to reflect that so many recoveries are performed within a depth range roughly equivalent to a walk across a large room, but that is typically the case. Perhaps no other enterprise besides ours, public safety diving and diving in general, better exemplifies one of mankind's most troubling limitations, that something can be so near, yet so far away.

Exposure Protection

Wet Suits

Whether they dive in Prince William Sound or off the Florida Keys, public safety divers need some type of exposure protection. Wet suits help keep divers warm by trapping and insulating a layer of water next to the diver's skin. The wet suit itself is a spongy material with thousands of gas bubbles inside. These bubbles act as insulation, trapping body heat. The thicker the suit, the greater the insulation. One of the drawbacks of the bubbles, however, is that they compress with increases of depth. The smaller the bubbles become, the less they're able to provide insulation. Thus, with a wet suit, a diver's thermal protection decreases with depth.

For public safety divers, wet suits have other drawbacks as well. They shouldn't be used in water colder than 50°F, and so they aren't appropriate for year-round use in colder climates. Because wet suits allow water to come in contact with a diver's skin, they offer no protection whatsoever against hazardous materials. Petroleum products such as gasoline from a submerged vehicle or creosote from pier pilings will break down the neoprene, causing the suit to disintegrate. Repulsive as it sounds, wet suits also tend to absorb body fluids and tissues from corpses during body recoveries. Once a wet suit is contaminated, it's virtually impossible to clean. Thus, if you use a wet suit, you should use it with caution, and only in pristine bodies of water.

If your team is unable to afford dry suits, be sure that the wet suits you use fit properly and prevent too much water exchange, and that they don't restrict breathing or movement. The thickness of the suit you need will vary, depending on the temperature of the water. Also, be sure that your SOGs state that you won't enter the water contaminated by petroleum products, sewage, stagnation, or other chemical or biological hazards.

Wet suits must be hung to dry after use. Because the neoprene will crease, wet suits must be kept on oversize hangers, which may create a storage problem.

Dry Suits

A dry suit encapsulates all parts of a diver's body except his hands and head, where wrist and neck seals prevent water from entering. Surrounded by a layer of air within, the diver stays completely dry and is able to wear warm, insulative underwear beneath the suit. To compensate for the compression of the air in the suit as the diver descends, dry suits have a power inflator, much like a buoyancy compensator device. These special characteristics mean that a dry suit, unlike a wet suit, is a unique piece of equipment that requires special training and certification. Still, dry suits offer many advantages to public safety divers, including greater thermal protection, the ability to dress quickly, ease of storage, and protection from hazardous materials.

A dry suit may be made of neoprene or crushed neoprene; trilaminent; or vulcanized rubber. Neoprene provides the greatest insulation, but vulcanized rubber is often the material of choice for haz mat concerns. Suits of vulcanized rubber also require less weight compensation. Whatever type you choose, make sure that the manufacturer can provide documentation of having tested it with various hazardous materials.

Unlike a wet suit, a dry suit of vulcanized rubber never needs to be hung. Simply

Dry suit and harness.

dry it with a towel, powder its seals, wax the zipper with paraffin, and the suit is ready for storage. A dry suit can be rolled tightly and fitted into a small duffel bag, which is an important consideration when gear must be packed neatly and efficiently into a compartment on an apparatus. Use talcum powder or cornstarch when treating the seals, and avoid using any powders with fragrance, since the chemicals in them may cause premature dry rot. Paraffin wax is the preferred lubricant for zippers, since beeswax and commercial dry suit waxes are too sticky, and the dirt that clings to them will damage the zipper. Do not use petroleum-based products, since these are incompatible with rubber and will destroy the zipper. You should wax the zipper on a dry suit after every use. Close the zipper and rub the paraffin block along the teeth. Unzip the suit before storing it.

Hoods

The most important function of a hood is to reduce heat loss. Approximately 25 percent of body heat is lost through the head, so wearing a hood can make the difference between staying functional and getting cold. In water of less than 70°F, the hood should be a minimum of three millimeters thick to decrease heat loss and help prevent vertigo induced by having cold water in the ears. Ideally, the hood should be rubber-covered, rather than lycra-covered, to reduce evaporative heat loss when the diver surfaces and has his head out of the water. Some dry suits are available with built-in hoods. If you're using a suit with a built-in latex hood, note that you need to wear an insulative skull cap under it to gain any thermal protection. Skull caps are available from dry-suit manufacturers. A dry suit offers its best protection against hazardous materials if it has a built-in hood that seals around the diver's face and under a full-face mask.

A wet-suit hood should have vents to prevent the buildup of air under the hood, allowing the diver to descend more easily and reducing neck strain from fighting buoyancy that is trying to pull the head up. A hood should have two vents, one at the top to vent air when the diver is upright, and another at the back of the head, to vent air when he's in a horizontal position, as during search. Vents can be made with a small, simple X cut.

Divers who use a full-face mask may wish to use a hood specifically made to go with a full-face mask. Such hoods have a ring of smooth, thin rubber that encircles the diver's face, allowing for a better seal.

Protective Helmets

In certain situations, such as diving below an overhead obstacle or in moving water, head protection is a good idea. Divers should use a light, plastic helmet, ideally worn over a hood. You'll probably have to remove the padding inside the helmet, since it tends to float, but the hood itself will act as padding for the helmet. Cave-diving helmets are an excellent choice, since they're specifically made for diving and you can mount lights or a camera on them.

Gloves

We lose about 15 percent of our body heat through our hands, and we search by running our hands over the debris-strewn bottom. Thus, gloves are important both to retain heat and to protect against injury. Because some environments have so many hazards, it may be necessary to wear leather work gloves over thin neoprene or dry-suit gloves for better protection. Kevlar® glove liners can be used when there is a potential for contacting sharp debris. Dry gloves are highly recommended for diving in contaminated water.

Boots

As with our hands, we lose about 15 percent of our body heat through our feet, so insulating boots or dry-suit socks, such as wool or fleece, are also a necessity. Whether built into a dry suit or worn with a wet suit, the boots should be hard-soled to protect the diver's feet around the dive site.

Harnesses

A diving harness is designed to allow a tethered diver to have both hands continuously free. Properly worn, harnesses also permit a diver to maintain a taut line without any discomfort or effort, as well as to feel signals through perhaps 150 feet of line. Because the attachment point is at a constant location, a diver can monitor his ascent rate by the change in the angle of the line across his chest. The angle of the line will also give early indication if the tether becomes snagged.

A good diving and water rescue harness should sit across the solar plexus so that it won't interfere with the diaphragm, respiration, or the lower ribs. It should have

Attachment point at harness.

adjustable shoulder straps so that the chest strap can be adjusted to the correct position. Feed the ends of the shoulder straps back through their adjusting rings and tape them down so that they lock in place under stress. The stitching of the harness should be strong enough to carry the weight of two adults, but it doesn't need the drop strength or the leg straps of rappelling harnesses. The webbing should be stiff enough that it won't roll on itself, bind up, and become ropelike. If the webbing does roll, it can lie between two ribs and push on them, causing discomfort when tension is applied to the harness. An X design on the back will prevent unnecessary pressure on the cervical region of the spine.

Although rappelling harnesses may seem to be suitable for water operations, they shouldn't be used for public safety diving. Rappelling harnesses have a low tether point, which will put the wearer in a vertical position in the water. For the same reason, tether lines shouldn't simply be tied around a diver's waist. A tether that's simply tied on can also rotate around the body, so the diver may have to search to find it and he won't be able to feel snags or signals as easily. The practice of holding a tether line in hand, of course, introduces certain problems, since doing so leaves only one hand free, and the backup diver will have no way of finding his way to the primary if the latter drops the line. The practice of holding a tether in one hand also leads to inaccurate search patterns due to changing positions of the hand with respect to the body. Some public safety divers and trainees balk at wearing a harness, believing that it presents an entrapment hazard. However, if any part of a harness system can entrap a diver, it's the tether. If the primary diver's line becomes entangled, then the backup diver will clear it and make sure that the diver is clear as well. If the tether line is too entangled to clear, then the backup diver will clip a contingency line to the primary diver's harness, cut the original tether, and take the primary diver to the surface. Far from being a hazard, a harness is an essential piece of equipment for public safety divers.

None of the above statements offer any argument for making public safety dives in black water without a tether. It is a myth that a backup diver will be able to follow a trail of bubbles in anything other than still water with good visibility, and if the primary diver has stopped breathing, there will be no trail to follow. In one common contingency plan, a primary diver in distress will release a tethered buoy, but this method has several flaws, not the least of which is the entanglement hazard that it presents to the backup. Commercial divers work on tethers for good reason, and so should public safety divers in conditions of limited visibility.

One commonly taught procedure is to use a harness tether, plus an added loop in the line for a handhold. This is not only useless and ineffective, it can also be dangerous. It's safer than simply holding the line in one hand, in that the line can't be lost, but nearly all of the other drawbacks apply. If the loop isn't placed at the point where the arm is at full extension, then the diver will have to rely mainly on the strength of his arm to keep the line taut. Any slack that develops will increase the risk of entanglement and decrease the accuracy of the pattern. Divers who use this technique say that it helps them feel the line-pull signals, and it makes them more comfortable to have one hand on the tether. Still, a diver with a proper harness should have no trouble feeling line-pull signals in his body, and having two hands free will allow him to perform a search pattern faster and with greater accuracy.

To tether a diver, secure a line to the front D ring of the harness with a figure-8 knot and a locking carabiner. The D-ring tether point can sit slightly off-center to the solar plexus so that when the diver is in the proper search position, the line won't end up between his legs. Teams may wish to purchase harnesses that also have a D-ring tether point on the back, allowing the harness to be used for tethering tenders to steep embankments and for other surface operations.

For overhead-environment dives, such as underwater vehicle operations, use duct tape to secure the lock of the carabiner. This will prevent rope abrasion and accidental opening of the gate. Also tape the figure-8 knot. Fold the tail of the tape into a tab so that you can find it readily when it comes time to remove it. Putting duct tape on a carabiner lock also reduces the chances that the carabiner will freeze shut during cold-weather operations.

Contingency Lines

A backup diver descending swiftly through murky water along the primary diver's tether line might become entangled in the same obstacle that snared the primary diver. Moreover, if he's groping his way along the line by hand, he'll have only one free hand available to manage whatever situation he encounters below. The solution is to use a five- to seven-inch contingency line, secured to the backup diver's harness carabiner and snapped onto the primary diver's tether line. One excellent setup is to use a plastic quick-release buckle in the middle, a carabiner or quick-release spinnaker shackle at one end to snap onto the line, and a brass ring at the other end, affixed to the backup's harness carabiner. Never attach yourself to someone else without having a quick-release capability.

A contingency strap offers some important advantages for a backup diver. Because it allows for a guided but hands-free descent along the tether line, it permits the backup, stressed by the urgency of the rescue scenario, to approach the primary diver without yanking on the primary's line. Pulling on his tether at this point might worsen the primary's plight or even promote injury. Naturally, having both hands free makes it easier for the backup to equalize and adjust buoyancy during the descent, as well as to assess and deal with the emergency. It can also save precious time, since if a backup were to lose contact with the tether line, the only sure way to find it again in low-visibility water would be to surface and pick it up again at a point nearer the tenders.

Contingency line snapped into harness.

Cutting Tools

Although many divers carry knives, as they were shown in their scuba courses, it's better if they carry one knife and at least two pairs of shears. Paramedic shears are among the most effective of tools to handle entanglement problems, and they don't pose the accidental cutting or stabbing hazard that knives do. Knives also aren't as versatile as shears, since to cut line, the line must be looped and the knife drawn through it—an action that requires two hands. Shears can cut through fishhooks, barbed wire, and many other materials, and they only require one hand to operate.

They're also cheaper than knives. If you do use knives at all, use only blunt-tipped ones, even if you have to grind down the tips. After using a knife in low-visibility or very cold water, ditch it. Do not try to resheathe a knife underwater. Divers have been known to cut themselves or their gear just trying to put their knives away. If you really need the knife back, you should be able to follow the profile map later to retrieve it, perhaps even as part of a drill.

Divers may also wish to carry other cutting tools as well, such as sturdy wire cutters or seat belt cutters. If you do carry specialized tools, carry them in addition to shears, rather than as substitutes.

Besides carrying the right type of cutting tool, you also need to have enough of them. Because there's always a chance that you'll drop a tool in black water and be unable to locate it, you should carry a minimum of three cutting tools. At least two of them should be mounted on the harness rather than the buoyancy compensator device. That way, if you need to discard your BCD, you'll still have cutting tools with you in a known, easily reached location. Do no mount cutting tools on your legs, where they will be far from reach and can become entanglement hazards as you kick and stir up debris, such as fishing line. Discarded weight belts have also caught on leg-mounted knives, resulting in fatality and injury. Tools should be mounted within the golden triangle; that is, the area from the mouth down to the bottom of the rib cage on both sides of the diver. Any equipment within this area will be easily accessible using either hand, increasing the chances for a self-rescue and facilitating the rescue of another diver. As always, each tool should be mounted in the same location every time so that the diver knows immediately where it is without having to look or grope for it.

Weight Belts

The weight belt is one of the simplest, least expensive pieces of gear that the public safety diver will use. It is also one of the most ignored, and one of the most deadly items worn in the water. Most dead divers are found wearing their weight belts. In many cases, divers struggle at the surface before drowning. If only they had ditched their weight belts, many of them might still be alive today.

One very important rule is usually ignored because divers don't fully understand why it's so important. Watch divers at any dive site and you'll see. Typically, the first pieces of gear that they remove at the surface are the mask and regulator, followed by the fins. After exiting the water, most divers remove their BCD. Finally, some divers remove their weight belt, while others continue to wear it during their entire surface interval, sometimes even while they demobilize.

Why is the weight belt typically removed last? Ask divers, and most will tell you that they hadn't really thought about it; that is to say, they weren't really consciously aware of having it on. If they're not consciously aware of having it on, there's little or no chance that they'll think about it when they're panicked or stressed. The longer they wear their belt on land, the more it becomes part of them, and the less likely it is that they'll ever think about dropping it to save themselves in the water. Also, divers who walk around with weight belts on are at greatly increased risk of drowning should they accidentally end up in the water. Sadly, this has been proved more than once. Many divers don't even think to ditch their belts in such situations.

Of those who do remember, most will shed the weight belt by opening the buckle and letting the belt fall away behind them. By this method, there is a real risk that the belt will get caught between the BCD and the diver's back; on the tank boot; or on a leg, knife, or fin. Other divers say that it's easier to take the belt off when the BCD assembly has already been removed. Consider such a diver struggling at the surface with an inflated BCD. Even if instinct tells him to drop the belt, the chances that he'll physically be able to are very small, since performing the same maneuver with an empty BCD on land is too cumbersome.

It's important to learn how to drop the weight belt properly, pulling it out and away from your body with your right hand before dropping it. Your left hand should simultaneously inflate the BCD. Practicing the proper technique throughout training and at the end of every dive will help reinforce it as a reflexive action in an actual emergency situation.

For public safety divers, the webbing of the weight belt should extend ten to twelve inches beyond the buckle. This is long enough to grasp quickly if the buckle opens accidentally underwater, but it's also short enough not to pose a snagging hazard. Never tuck the free end of the belt below the webbing, since this will negate its quick-release characteristics. We recommend that teams purchase a 150-foot roll of webbing so that they can make their own correctly sized weight belts. When donning a belt, make sure that the buckle is placed properly on the webbing so that it can close fully and properly. Metal buckles are recommended.

Weights should be symmetrically placed on the belt, not only for buoyancy control, but also for reasons of buckle placement. If the weights are asymmetrically placed, there's more of a chance that the belt will rotate, putting the buckle off to one side, where it'll be less accessible in an emergency situation. The belt can be offset by one to two pounds, though, to compensate for a nonbuoyant pony bottle. Weights should not be stacked on a belt, since the extra protrusion can make the belt difficult to ditch. Similarly, the BCD cummerbund should never impinge on the belt. Suit-compression compensators are highly recommended to prevent the weight belt from loosening and rotating as the diver descends.

When donning a weight belt, you should always orient it for a right-hand release. This is important even for left-handed people. Unless you have a right-mounted BCD power inflator, you need your left hand free to inflate the BCD, leaving your right hand free to ditch the belt once you reach the surface. It's also necessary from the standpoint of standardized rescue procedures, so that any diver can, by reflex, perform the ditching maneuver on any other diver. Never tuck the free webbing back under the belt, since this will make it less than a quick-release system.

Public safety divers should be overweighted by about two pounds to allow them to search the bottom effectively. Searches in moving water may require even more weight, depending on the speed of the water. However, bear in mind that most divers are already overweighted by about five to ten pounds. Take the time to do a proper buoyancy check during training sessions, and be sure to record the weight you need in your logbook. Adjust your weighting anytime you use a different exposure suit or if your dives occur in water with varying degrees of salinity.

On an actual call, even if you have conducted proper buoyancy checks, you may suddenly find that you're unable to descend. The harder you try to go under, the more buoyant you seem to get. What's probably happening is that the adrenaline is running high within you; you're breathing heavily, and that extra air is increasing your buoyancy by several pounds. The solution is not to go back to shore to don more lead. Doing so will only put you in a dangerous situation and delay the operation further. Instead, stop, focus, and slow your rate of breathing. Once the volume of air in your lungs has subsided, you'll be able to slip underwater.

Weight Harnesses

An increasing number of public safety divers use integrated weighting systems, in which weights are placed in special holsters in the BCD. These holsters have some sort of pull system that supposedly allows you to dump the weights in an emergency. Although the idea of not having to wear a heavy weight belt might sound appealing, integrated weight systems present some problems for both sport and public safety divers. Probably the most severe of these is that a diver wearing an integrated weighting system has a slimmer chance of being successfully rescued than one wearing a conventional weight belt, since there's a good chance that the rescuer won't realize the weights are integrated; won't know how to ditch the integrated weights; won't be able to ditch the weights because the victim is on his back; and may not even see the weight-dump pull tags. Also, every type of integrated-weight BCD has a different kind of pull system, further complicating the issue. Those who use integrated weights are typically less likely to adjust their weights to maintain proper buoyancy when changing exposure suits, since it's tedious to do with some systems. Not only is a BCD with integrated weights more cumbersome to handle on land, it also won't allow the wearer to put the weight in the most effective locations around his body. Moreover, dive coordinators and safety officers can't tell at a glance how much weight a diver is wearing. If your department does have integrated weights, put at least half of the weights on a weight belt, which can be removed at the surface to keep you afloat, can be removed at the end of each dive for practice and safety, and can help prevent an uncontrolled ascent should you ever need to ditch the BCD assembly underwater.

Ankle Weights

Ankle weights are important to have at the dive site for two reasons. First, they can be used to add weight easily to a diver who is unable to descend, a not-uncommon problem at a rescue scene. Simply clip an ankle weight around the taut tether line and send it out with a hard push to the diver. The diver can then clip the weight around his weight belt or BCD strap. The weights can also be clipped around the diver's tank valves, where they'll be out of the way and not lost when he removes his weight belt. Second, ankle weights may be necessary to keep a diver's feet and fins on the bottom. Dry suits, thick booties, and even plastic fins will float a diver's feet and legs off the bottom. Small ankle weights correct that problem.

Scuba Equipment

Entanglement is the number one concern for public safety divers in waters of limited visibility. Any diver's gear, especially that of a public safety diver, should be trim, with nothing dangling. Keep all hoses close to the body. Run the gauge hose under the diver's arm between the buoyancy control device and the body. If the gauge hose is merely secured to the outside of the BCD, it can be pulled far enough away for a snag or entanglement to occur between the hose and the BCD. This is not an easy place to reach. With the hose snug under the BCD, the gauges can also be secured by quick-release plastic clips to the BCD. However, don't secure gauges with metal clips that cannot be broken or cut. If a diver needs to be rescued and have his gear removed, his rescuer will have to unclip the gauges. If the rescuer cannot release the clip or doesn't see it, he should be able to break or cut it.

To secure inflator hoses, pass the hose from the regulator under the arm, with the BCD hose in its normal position over the shoulder. This position will hold the hose in place, but it will still allow a diver to vent his BCD when he raises his arms. If you use this setup, note that a backup diver will need to disconnect the inflator hoses to strip the BCD from an injured diver, a procedure that should be practiced. Short inflator hoses are available that will help keep the hose closer to the diver's body. Inflator hoses can also be secured under surgical tubing to prevent dragging. Make sure that the pony regulator hose runs snugly downward along the back of the BCD and under the diver's arm, where it can be secured with a mud mouthpiece and snorkel keeper in the golden triangle.

Buoyancy Control Device

A buoyancy control device is an inflatable vest or backpack that allows a diver to add air to change his buoyancy. Public safety divers should use jacket-style BCDs rather than back-flotation BCDs, since the latter were designed for cave diving and wreck penetration, and they'll force you face-down if you're carrying a body on the surface. Also, they often lack the pockets that public safety divers need for tools. A horsecollar BCD will float you face up, but it may make it difficult to transport a drowning victim properly along the surface. Make sure that all crotch straps are secured under the weight belt.

A buoyancy control device used for public safety diving should provide at least thirty-five pounds of lift. In addition to being two pounds overweighted to help him search the bottom, the diver may have to support the weight of a large adult, typically falling within a range of about eight to sixteen pounds in the water. At the surface, the BCD also has to support the weight of the diver's head, neck, tank valve, regulator, mask, and perhaps a hand or two. The BCD should be well constructed so as to withstand continual contact with the debris-strewn bottom, and it should have quick-release shoulder straps for one-handed removal. A bright color, such as orange, is an advantage for a rescue team. Multiple pockets for tools are essential. Avoid BCDs with lots of clips, D-rings, or other unnecessary appurtenances that would increase the risk of entanglement. Make sure that the BCD won't cover the dry-suit inflator, weight belt release, or harness tether point. One very useful feature found on specially made public safety diving BCDs is a built-in quick-release pony bottle pocket. The BCD should be of a hard backpack design, rather than soft, since a hard pack will keep the tank in one place and doesn't require extra lead to sink. Always make sure that device fits properly. If the team doesn't have one to fit you, don't dive until one is made available.

A buoyancy control device should always be worn with a dry suit. Although dry suits can be used to adjust buoyancy underwater, it isn't recommended that you do so. A BCD provides more flotation for a diver on the surface, which could be important in assisting another diver or holding a victim afloat. Wearing a BCD with a dry suit is an industry standard, and an officer who allows a dry-suit diver in the water without a BCD could face liability issues in the event of a death or injury. Consider what would happen if the dry suit were to flood. Even ditching the weight belt might not keep the diver at the surface.

Power Inflator

Every BCD should be used with a functional power inflator. If a power inflator malfunctions while underwater, the dive should be aborted for two reasons. First, a malfunctioning inflator could cause accidental inflation of the BCD, possibly resulting in a dangerous rapid ascent. Second, a diver shouldn't orally inflate his BCD, since he would then risk ingesting contaminated water. A diver wearing a full-face mask cannot inflate his BCD without first removing his mask.

The power inflator should not have an integrated alternate second stage. These units tend to collect sand and mud, leading to frequent malfunctions. If you use an integrated octopus-type power inflator for sport diving, your primary regulator hose must be replaced with an octopus length of hose so as to share air effectively. The integrated second stage is designed so that you give up your primary and you take the integrated second stage. The safety hazards presented by this procedure are numerous.

Cylinders

Scuba tanks used for public safety diving should have a minimum capacity of eighty cubic feet, otherwise written as 80 ft^3. If the cylinders have J valves, the reserve mechanism should be deactivated. Tank boots aren't recommended for flat-bottomed tanks, because they can snag a dropped weight belt and promote damage to the tank bottom by trapping corrosion-promoting water.

Some dive stores have convinced teams that they should only use steel tanks with DIN fittings. In reality, those tanks are far more than most teams need. Aluminum tanks with K valves are perfectly suitable, as well as cheaper. However, if a team can afford it without cutting corners in other areas, the DIN system may be preferred. With a DIN system, the first stage of the regulator screws into the tank valve, so the DIN fittings are more durable and can withstand abuse that would cause a standard regulator to be dislodged from a standard yoke fitting.

Be sure that all cylinders have a current, valid visual inspection sticker and a current, valid hydrostatic test date. Also, whether you fill the tanks at the dive store or from your own compressor, be sure that the air is free of contaminants and that the fill station is tested regularly. The dive store should be able to provide you with documentation. The maintenance of your own unit is up to you. It is usually cost-effective to have a member of the department certified to perform the annual visual inspection.

Contingency Cylinder

In addition to the tanks worn by the divers, there should be a contingency tank on shore within easy access of the backup tender. A contingency tank is a designated emergency scuba tank set aside to be taken to a diver trapped underwater after he has been given the backup diver's pony bottle. This tank should have a regulator attached, a carrying handle, at least one extra cutting tool, a carabiner, and a submersible flasher. On the scene, it must be full, and the backup tender should check its status before the primary diver enters the water.

If the contingency tank is a standard aluminum 80-$ft.^3$ tank, be aware that it will float when some of its air is used. Since the last thing a diver needs while being cut from an entanglement is to wrestle with and possibly lose the tank from which he is breathing, aluminum contingency tanks should be weighted with about two pounds of lead about eight inches above the bottom.

Pony Bottle

Imagine that you become entangled on the bottom toward the end of a search dive. Unable to free yourself, you signal for the backup diver. Just as you do, your regulator starts to free-flow. Within thirty seconds, your tank is empty. You reach for your octopus and realize that it won't do you any good, since it's supplied by the same tank. Maybe the backup diver reaches you in time to give you his octopus, but now he can't reach around you to solve the entrapment problem. With both of you breathing off the same tank, you'll quickly use up the air. Does he leave you to drown, or does he stay until both of you run out of air and drown?

Public safety divers should use pony bottles. These small, ancillary tanks allow divers to have an air reserve even when their primary has been emptied. Having a pony bottle mounted to the main tank with a quick-release harness gives a backup diver a supply of air that he can pass off to another diver, allowing him to leave and return with still more air. Pony bottles are essential for safe, responsible public safety diving.

At minimum, a pony bottle should be an 18-ft.^{3} tank with an independent regulator. For depths of sixty to a hundred feet, divers should use bottles of 30 ft.^{3} or larger. For low-visibility diving, depths of more than a hundred feet require surface-supplied gas and a bailout bottle.

Some divers think that wearing a pony bottle makes them more susceptible to entanglements. If entanglement is truly a concern, however, having the pony bottle only becomes that much more of a necessity, since you'll need a system that will allow for self-rescue and a diver-to-diver exchange of air. Besides, if worn properly, a pony bottle does little to increase the risk of entanglement.

As with the contingency tank, a contingency pony bottle with a regulator should be within easy reach of the backup tender. When passing off an aluminum bottle underwater, be sure to maintain control of it so that it doesn't float up and hit you in the face, or perhaps someone else farther up.

Recreational divers carry pony bottles as sling bottles at their sides. These divers are midwater divers. They typically don't crawl or swim along a debris-covered bottom in zero visibility. They also don't make continuous sweeping motions with their arms. Public safety divers are better off if their pony bottles are mounted away from bottom debris and the action of their arms. Securely mount the pony bottle up against the BCD back pouch on the main tank. There should be no space between the BCD and the pony bottle. A diver doesn't have to be able to release his pony bottle himself. Rather, an entangled primary diver will take the pony bottle off the backup diver's BCD assembly.

It's best if the pony bottle is mounted on the right side of the tank, so that the regulator second stage is properly set up for the diver to use. Wherever you mount your pony bottle, be sure that it has some type of quick-release system. The mounts can either be of webbing harness, from which the bottom can simply be pulled out, or two metal brackets, one mounted to the pony bottle and one mounted to either the main tank or the BCD strap. Metal, bracket-type pony mounts are ideal, since they'll hold the bottle most securely. However, they are quite expensive, sometimes costing as much as the bottle itself. Although webbing harnesses don't quite hold a pony bottle rigid, they are far less expensive and may therefore be a better option for a dive team.

There are four ways to pass off a pony bottle to another diver. The least expensive and perhaps most complicated method is to remove the regulator and switch to another regulator mouthpiece from the pony bottle. If the divers are in contaminated water, however, the pony bottle regulator will be immersed in the contamination, then inserted directly into the recipient's mouth. A diver wearing a full-face mask will have to remove the mask to make the transfer, exposing his mouth, nose, eyes, and skin to the same hazard. Even if the water isn't contaminated, exposing the face suddenly to cold water may cause gasping, laryngospasm, and possibly drowning.

The second option is to use a standard commercial block. This is a valve system that allows a diver to keep his regulator or full-face mask in place when he switches

It's best if the pony bottle is mounted on the right side of the tank.

to an alternate source of air. These are designed for surface-supplied commercial divers, however, not public safety scuba divers. Although commercial blocks typically cost well over $250, many divers choose to build them from scratch. Home-built blocks open up the builders to liability problems, however, in the event that the system fails or is involved in some type of mishap. The real problem with commercial blocks is that, although they do allow a diver to change gas sources, they don't usually allow a diver to pass a self-contained, redundant gas source to another diver. Such a system won't work for public safety operations, since the backup must be able to pass a pony bottle or other cylinder to an entangled primary.

Some full-face masks have ports for two second stages. The front port is for the primary mouthpiece, and the other is for a pony bottle or octopus mouthpiece. With this system, the diver in trouble can switch to a second air source without having to

Regulator-mounted diving block.

remove his mask, although the donor will have to remove his regulator.

The fourth and safest method is to use a regulator-mounted diving block designed to make the transfer from a main tank to an alternate gas source, such as a pony bottle or other cylinder. The unit is connected to both a main gas cylinder and a redundant gas supply, such as a pony bottle. When activated by removing a safety clip and pushing up on a valve, the unit shifts the diver's breathing source from the main tank to the pony bottle. Unlike normal commercial diving blocks, this device was specifically designed to address the needs of public safety divers. Note that some recreational trainers advise teams to go with double primary tanks rather than carry a pony bottle. It's true, a diver will have more gas that way, but instructors who offer such advice are forgetting the need to be able to pass off a bottle to a diver in distress.

Regulators

Although regulators don't necessarily have to be complex or expensive, a team should consider several options before making any purchases. All regulators should be of the downstream design; that is, in the event of a malfunction, they should free-flow. First-stage regulators for primary tanks should be equipped with a second-stage regulator, a power inflator hose, a submersible pressure gauge with an attached depth gauge, and a dry-suit inflator hose if a dry suit is worn. A balanced regulator that will deliver air at a consistent pressure as tank pressure falls is needed for the primary regulator. Entangled or stressed divers, low on air and with an unbalanced regulator, can quickly find themselves in trouble as breathing resistance becomes increasingly stronger.

One consideration that teams need to evaluate when purchasing regulators is the need for environmental protection. Environmentally sealed regulators offer two advantages. First, they reduce the chances of a regulator freeze-up during cold-weather operations. Second, they're easier to decontaminate and may offer fewer contamination hazards to a diver.

Teams that work in moving water may wish to purchase regulators with adjustable second stages. The adjustment on the second stage can be helpful for tuning down the regulator so as to reduce free-flows when facing the current. Teams that will be working in water moving faster than two knots may wish to purchase side-breathing regulators.

Unless a team plans on becoming trained and operational in deep diving, there is no point in spending money on expensive deep-diving regulators. Instead, most teams should seek inexpensive, environmentally protected, reliable regulators that are low-maintenance and easy to clean.

The pony regulator should only have a single second stage, and perhaps a tiny pressure gauge that screws directly into the first-stage high-pressure port. Removing the exhaust port tee from the second stage of the pony regulator will prevent it from scooping mud or dirt that could block the exhaust port, and it will prevent the tee from being mistaken for a mouthpiece in black water. Some second stages come without exhaust port tees. These are recommended. Also recommended are second stages that can be used inverted, such as side-breathing regulators. This allows the mouthpiece to go equally into the mouth of the wearer or another diver without having to bend the hose back.

The second stage of the pony bottle regulator should be secured within the golden triangle. At minimum, it should be secured with a mouthpiece cover that protects the entire mouthpiece and opening to prevent it from scooping up mud and debris while the diver crawls over the bottom. The second stage can be secured further by using a snorkel keeper around the mouthpiece as well.

Public safety divers should not wear an octopus. An octopus, or safe second, is an extra second-stage regulator, carried to provide another diver with air. This device is for shallow-water, high-visibility, buddy-system sport diving. An octopus isn't a true alternate air source; rather, it is merely a second mouthpiece, and it will do absolutely nothing for you if you use up your gas supply. Furthermore, in ice-diving operations, an octopus will significantly increase the chance of a free flow. Some divers are incorrectly taught that the octopus is a redundant second stage in the event that their primary second stage fails. If your downstream primary second stage fails, it can only free-flow. If this occurs, you don't want to switch to an octopus because that will increase the chance of a first-stage free flow, which could empty your tank very rapidly.

Gauges

All divers should be equipped with both a depth gauge and a submersible pressure gauge (SPG). Both the depth gauge and the SPG should be analog. Although it's acceptable to use a digital computer as a backup to the analog gauges, digital gauges shouldn't be used, since they're harder to read in low-visibility water.

Sometimes the water you're searching in is so murky that you can't read the numbers on analog gauges. In such instances, phosphorescent gauges may still allow you to see the position of the needle. If 1,000 psi is at the 10 o'clock position on your gauge, for example, all you'll have to remember is that you have to be home at 10 o'clock.

For those occasions when the water is so black that you can't read a gauge at all, even when you hold it right in front of your mask, there is an easy solution. Fill a sealable plastic bag with clear water, then tape it on top of your gauges. When you press the bag to your mask and shine a small light into it, you'll be able to read the gauges even in the worst of conditions.

Teams may also wish to equip pony bottle regulators and contingency tanks with miniature SPGs that screw in the first stage of the regulator. Although a diver won't be able to read these gauges underwater, his tender can check them during the dressing phase to ensure that the pony bottles are full. Otherwise, the tender would have to check the pony bottle with a separate pressure gauge.

COMPUTERS

Although there's nothing wrong with using dive computers in public safety diving, they're generally unnecessary for most teams. The dive times of rescue/recovery missions are generally limited to twenty-five minutes, with maximum depths of fifty to sixty feet. Since the true advantage of a dive computer is demonstrated in multilevel dives of longer duration, they have little use in our operations other than to record the maximum depth and time on the bottom. For public safety diving, leave your expensive computer at home, unless you are performing deep or long-duration operations.

SNORKELS

A snorkel is the single most useless piece of equipment for public safety diving. It beats against your head in a current, it gets entangled, and it can dislodge the mask. Also, putting a snorkel into your mouth allows you to aspirate and ingest contamination from polluted water, and it increases the chances of aspirating water. At no time should a public safety diver switch from a regulator to a snorkel.

MASKS

A standard face mask should be sized to fit the individual diver. A department should not go out and purchase ten of the same mask, unless every diver has the same-shaped face. To size a mask, gently lay it on the diver's face and look around the skirt to make sure that it touches at every spot, without any gaps. If you only use the common procedure of sucking in through your nose to see whether the mask will stick on your face, you may end up with a leaking mask.

Be sure to clean the inside face plate regularly to prevent the mask from fogging. Use mask cleaner, paste, toothpaste, or a soft abrasive cleanser. Although fogging makes no difference whatsoever in black water, it will increase the chances that a diver will remove his mask at the surface. Divers should keep their masks on until they are out of the water. If you use standard masks, be sure to wear the strap under the hood to prevent loss of the mask if it becomes dislodged. Also, divers may wish to replace their silicone mask straps with neoprene straps. In addition to being more comfortable, neoprene straps offer a few practical advantages. They're easier to don, easy to adjust, and they rarely break.

Standard face masks are fine for public safety divers who operate in clean, relatively warm water with enough visibility to make buddy diving effective. Standard face masks and regulators are not acceptable for contaminated water, however. Because they keep the face warm, allow for communications, and help protect against hazardous materials, full-face masks are excellent for public safety divers.

A full-face mask can be either a demand mask, in which air is supplied at ambient pressure (like a normal regulator), or a positive-pressure mask, in which air is supplied at a pressure greater than ambient pressure. For public safety divers, positive-pressure masks are the better option, since the positive pressure inside the mask prevents water from leaking in, thus reducing the potential for contact with contaminants. Positive-pressure, full-face masks also increase the chances that an unconscious diver's airways will stay dry, which could make the difference between life and death. Both demand masks and positive-pressure masks provide a degree of thermal protection, and they allow for the use

Divers using a full-face mask must practice removing the mask underwater and switching to a pony regulator until it becomes a skill they can do by reflex.

of communication systems. One disadvantage of positive-pressure masks is that, without the correct hood, they're difficult to seal around the face without having a constant stream of bubbles emanating from under the mask skirt, or without causing the diver's hood to inflate with air. Also, a positive-pressure mask will cause you to use and lose air more quickly than you would with a standard regulator or even with a demand mask.

Although they're the better choice, remember that full-face masks require a higher degree of training. They must also be cleaned at the end of every diving day, and they generally require more maintenance. Divers who are new to using full-face masks, especially positive-pressure ones, often increase their rate of air consumption by approximately 20 percent, which should be taken into account.

Historically, one of the disadvantages of full-face masks in general has been that, whenever divers wore them, they would be breathing from their main tanks. Backup divers stationed topside couldn't have their masks fully in place and be ready to go without depleting a significant amount of air from their tanks. However, vents are now available that will allow a diver to breathe surface air. When he must descend, he simply pushes in the vent and automatically switches to air from his tanks. Such vents are available as aftermarket items that can be added to a full-face mask.

Ideally, the surface vent should be mounted in the oronasal pocket of a full-face mask. There are four benefits to placing it there. First, by being in line with the airway, the vent allows for easy breathing. Second, the diver must only exchange the dead-air volume of the oronasal pocket rather than of the entire mask, thereby reducing the buildup of carbon dioxide as he waits at the surface. Third, regulator free-flows can be managed by partially pulling out the vent, allowing excess air to flow out of the vent and permitting the diver to leave the mask in place. Fourth, a vent mounted low on the mask has a smaller chance of becoming entangled.

Fins

Commercial divers walk to perform their jobs. Public safety divers do not. Those who operate in high-visibility waters either swim or are towed above the bottom; those who operate in low-visibility waters search along the bottom, swimming in controlled sweeps by means of a tether. For these divers, fins may be considered an absolute necessity. Because fins are an industry standard, department officers who use a no-fins policy could open themselves up to serious liability issues should a death or injury occur during operations.

Wrap the outside strap and buckle of each fin with duct tape to decrease the risk of entanglement. Keep the inside strap free if the fin requires adjustment. Trying to adjust or disentangle an outside strap can result in painful leg cramps, whereas inside straps are easy to reach if the legs are crossed.

Some of the newer plastic fins have a tendency to float, which is problematic for public safety divers. Those who operate in very shallow water may find that buoyant booties and fins raise their legs so high that it becomes difficult to search effectively along the bottom. The result is that divers compensate by overweighting their belts. Divers who operate in black water need to keep their fins on the bottom as they search with their entire body. Having buoyant feet can be a hazard to dry-suit divers, since it poses the risk of accidental inverted ascents. Although rubber fins are preferred, if your fins float, you may need to wear a pair of ankle weights to keep your feet where they need to be.

Surface-Supplied Diving Equipment

By connecting a second stage to a topside compressor or a bank of cylinders, surface-supplied equipment allows a diver to receive unlimited quantities of air. It allows a diver to operate extensively and to exert himself with far less concern about running out of air. For this reason, surface-supplied air is highly recommended for dives to depths of a hundred feet or more, where high air consumption is a problem. It's also recommended for overhead environments known to be contaminated and areas full of potential entanglements.

Surface-supplied diving requires more physical effort on the part of the diver. It also requires equipment very different from that of normal scuba gear. In addition to the surface compressors, panels, and breathing equipment itself, the diver must wear bailout bottles in case of a malfunction in the main line, and backup divers with fully redundant sources of air are still necessary. Surface-supplied divers often wear a full diving helmet, which offers even greater encapsulation protection than a full-face mask.

Surface-supplied systems are expensive, and they tend to require more maintenance. Considerable training is required to use and maintain it. The emergency procedures differ from those used for standard scuba gear, and tenders must be trained in those procedures as well as divers. Some teams opt to use surface-supplied systems because they're trained by commercial divers. However, if your normal range of operation is less than fifty feet, severe contamination is not a concern, and the entanglement threats are not extreme, such a system would probably be expensive overkill.

Surface Signaling Equipment

Although they operate on tethers to shore, many divers should still carry equipment to allow them to signal from the surface in case they become disconnected. A waterproof, submersible flasher should be used for low-light or night operations. The flasher will allow shore personnel to know a diver's location on the surface at all times, and it will alert them if he surfaces unexpectedly.

Divers should also carry a whistle or other audible signaling device. Whistles should be pea-less, since cork peas tend to rot or compress in the water. If you work in cold climates, avoid using metal whistles, which will stick to the lips when temperatures are low. A better option is to use an air-horn system that connects into the power inflator hose. Besides providing a louder signal than a whistle, the diver will be able to sound an alarm without having to remove his regulator or put a contaminated whistle in his mouth.

Dive Lights

Many teams assume that if they conduct night or low-light operations, they'll need dive lights. Before spending money on expensive lights and batteries, however, consider carefully whether you really need them. If you dive in black water, lights will do you no good whatsoever, other than to help you read your gauges. Visibility in black water is limited by the particles in suspension, not just a lack of light.

If you dive in high-visibility water, dive lights are valuable for night searches. Those that can be worn on the back of the hand or mounted as headlights are particularly useful, since they leave both hands free. If you do use lights for a night operation, be sure that you have plenty of batteries on hand.

Mixed Gases

The nitrox mixtures used in diving have more oxygen, and therefore less nitrogen, than standard air. The air we breathe at the surface is composed of about 21 percent oxygen and 78 percent nitrogen. The two most common mixtures for sport diving are Nitrox 32, containing 32 percent oxygen and 68 percent nitrogen, and Nitrox 36, composed of 36 percent oxygen and 64 percent nitrogen.

The term "enriched air" is often used interchangeably with nitrox, although it specifically means any gas mixture containing a higher percentage of oxygen than standard air. Enriched air is often abbreviated as EAN_x. Thus, Nitrox 32 could also be written as EAN_{32}.

Enriched air has three common benefits for divers, all of them accruing from the lowered proportions of nitrogen. The first benefit is that it can decrease the risk of decompression sickness if the diver stays within the no-decompression limits created for diving with standard air. Second, it can allow for longer dive times than would be safe using standard air. Third, enriched air has some value in decreasing the effects of nitrogen narcosis.

Many people believe that enriched air is used for deep diving. In reality, the increased oxygen content of enriched air means that, because of potential oxygen toxicity caused by increased pressures at depth, it shouldn't be used at great depths. In fact, Nitrox 36 is considered unsafe if breathed below 110 feet. Nitrox should never be used where a diver could accidentally end up at a deeper depth than planned, especially in low-visibility water, where a diver might not even know how deep he is.

Most search and recovery dives occur in water shallower than fifty feet, where air consumption, rather than dive time, tends to be the limiting factor. Given the short rotation times of public safety divers, the longer bottom times permitted by nitrox are useless. Additionally, nitrogen narcosis isn't a problem at those depths, nor is the risk of decompression sickness. Although many proponents of enriched air state that it reduces post-dive fatigue, there is no evidence or research to support these claims. Issues of expense and high maintenance requirements also preclude using it for rescue/recovery. Finally, the use of nitrox requires specialized training and certification beyond that of a normal open-water scuba course and a Level I public safety diver specialty course. Because of the dangers of uneducated divers using the wrong equipment, if a team uses nitrox, all of its members must be certified.

In reality, it's difficult enough to get public safety divers to do at least twelve standard search dives annually. What, then, are the chances of adding annual training and drills to make sure that all divers, tenders, surface personnel, and tank fillers are safely trained for nitrox operations? The training alone for all of these personnel, plus periodic refresher training, can cost thousands of dollars. Add in the cost of nitrox-capable tanks and regulators, as well as at least two gas-mixture analyzers, and the expense can quickly become prohibitive.

Given these considerations, the only public safety teams that should consider using enriched air are those with large budgets and expansive training schedules, and that commonly dive to depths of sixty to a hundred feet, where the greatest benefits of enriched air are to be realized. Those divers should use standard air-table profiles. Once dive times are extended with nitrox tables, the safety benefit of using nitrox is gone.

Many divers are also familiar with heliox and trimix, which are highly specialized gas mixtures used for deep technical diving. Heliox is a mixture of helium and oxygen, and trimix is a mixture of helium, nitrogen, and oxygen. Because they are used specifically for deep diving, these gases have little use for most public safety operations. If divers need to descend beyond a hundred feet, they should be using surface-supplied gases. Furthermore, the equipment and training needed for heliox and trimix are both extensive and expensive.

It should be noted that mixed-gas diving for public safety purposes may fall under OSHA regulations.

Prioritizing Purchases

One of the realities of public safety diving is that every team is limited by budget. The majority of teams can't afford to purchase every piece of gear that they want right away, so they must assemble their arsenal of equipment piecemeal. Although this can be a frustrating process, teams should first purchase those items that are essential for keeping all of their members as safe as possible.

After a team has purchased its basic scuba gear, its members should next be outfitted with personal flotation devices, without which the team should not be operational. For the same reason, the next priority is to acquire pony bottles and regulators. After these essential elements are in place, the team may purchase positive-pressure full-face masks to protect divers from cold or contaminated water, as well as to keep the airway dry if the diver becomes unconscious. Depending on the quality of the water in the response area, dry suits may be the next item bought. Communications systems generally fall next in the order of priority. Although many teams rank communications systems above dry suits or even pony bottles, a communications system is of no value in helping a diver trapped on the bottom without air or in keeping him warm and protected from contamination. Because communication can be accomplished by using line signals, communication systems aren't as important as equipment that directly serves physiology and the preservation of life.

Still, a communications system should be a priority item for overhead-environment diving; where direct-line tether access can be lost; where tenders can't easily monitor a diver's breathing rate by watching the bubbles; and whenever divers aren't tethered. Such a system is also important for waters that are contaminated, moving, or of limited visibility.

Standardization of Equipment

Let it be said that, no matter what type of equipment your team chooses, it should all be standardized. Ideally, all dive gear should be of the same type and set up the same way. Standardization allows for greater flexibility when divers have to share their gear. It also means that each diver is familiar with everyone else's. A diver can easily remove a pony bottle from another diver's pony holder if the holder is the same as his own, but he may never be able to figure out how to remove a pony from an unfamiliar bracket in black water.

For many teams, the equipment on which an operation is based is supplied by individual team members using their own personal gear. Because of differences in preference, the features of the gear can vary greatly. Even so, there can be some form of standardization. Pony bottles, for example, should all be worn on the same side. If the equipment isn't standardized, make sure that the primary divers, backups, and ninety-percent-ready divers all review each other's gear before commencing any operation. With their eyes closed, they should be able to remove each type of BCD from another diver, find the second stage of another diver's pony regulator; remove pony bottles from other divers; and ditch each others' weight belts. Training is the key, and such familiarization needs to take place long before the alarm ever sounds. Take the time to become familiar with every item used by the team.

Prepping Equipment

Time is of the essence in an emergency situation, and prepping dive equipment for the call will reduce the amount of time required to gear up for an oper-

The BCD, regulator, pony bottle, and pony bottle regulator should already be assembled to the tank, with the hoses and mouthpieces secure.

ation. In some instances, divers and tenders can reduce response time by dressing en route to the scene. In other instances, team members are unable to do so. For either scenario, steps can be taken before a call that will reduce both fatigue and response time.

To configure your equipment to meet the emergency, the BCD, regulator, pony bottle, and pony bottle regulator should already be assembled to the tank, with the hoses and mouthpieces secure. Cutting tools should be attached within the golden triangle, with at least one pair of shears on the BCD, and at least one pair of shears and a knife (or a backup set of shears) on the harness. The seals of a dry suit should be powdered, and its zippers waxed with paraffin. The top of the dry suit should be rolled down to the waist to allow for quick entry. Wet-suit farmer johns should be rolled down to the waist. Wet-suit tops should be slightly zipped up (if they don't have beaver tails) so that all the diver has to do is step in and finish zipping. The shoulders should be pulled back inside-out, but don't turn the sleeves inside-out.

Gloves and standard masks should be kept in the hood, and one boot can be stored in each fin. Each diver-tender team's equipment, minus the BCD-tank assembly, should be stored in one well-marked bag or bin in the response vehicle. Don't store all exposure suits in one place, PFDs in another, and dive gear in a third. Having to collect equipment from several different places results in downtime and lost gear. Pack the gear bags or bins in such an order that what will be needed first is on top.

Scuba gear is life-support equipment and, as such, it should be checked after each dive and thoroughly inspected on at least a monthly basis, regardless of use.

Study Questions

1. Wet suits shouldn't be used in water colder than ________.

2. What is the purpose of a power inflator in a dry suit?

3. What is the most important function of a hood?

4. Why shouldn't rappelling harnesses be used for water operations?

5. For a guided but hands-free descent along a tether line, a backup diver should always use a ________.

6. Although many divers carry knives, as they were shown in their scuba courses, it's better if they carry one knife and ________________________.

7. Where should tools be mounted?

8. What is the proper way to shed a weight belt?

9. When donning a weight belt, how should you always orient it?

10. To allow them to search the bottom effectively, public safety divers should be overweighted by about ________.

11. True or false: A diver wearing an integrated weighting system has a lesser chance of being successfully rescued compared with one wearing a conventional weight belt.

12. At minimum, a buoyancy control device used for public safety diving should provide how much lift?

13. Why shouldn't the power inflator have an integrated alternate second stage?

14. Scuba tanks used for public safety diving should have a capacity of at least how much?

15. True or false: Aluminum contingency tanks should be weighted with about two pounds of lead about eight inches above the bottom.

16. What are the minimum requirements for pony bottles used no deeper than sixty feet below the surface, as stated in the text?

17. All regulators should be of what design?

18. All public safety divers should be equipped with what two gauges?

19. True or false: For public safety divers, demand masks are preferred over positive-pressure masks, since they conserve air.

20. True or false: Because fins are an industry standard, department officers who use a no-fins policy open themselves up to serious liability issues should a death or injury occur during operations.

21. Surface-supplied air is highly recommended for dives to depths of ________ or more.

22. Although many people believe that enriched air is used for deep diving, Nitrox 36 is actually considered unsafe if breathed below ________.

23. True or false: Because of the dangers of uneducated divers using the wrong equipment, if a team uses nitrox, all of its members must be certified.

24. True or false: Mixed-gas diving for public safety purposes may fall under OSHA regulations.

Chapter 7

Cylinder Safety

Cylinders manufactured under the U.S. Department of Transportation 3AA and 3AL requirements are designed to accept many thousands of fills over their lifetime. Even so, damaged tanks resulted in eight cylinder explosions in 1997, most of which could have been prevented by careful visual examination.

When learning about their life-support systems, divers often aren't told about the potential hazards stored inside the cylinders. An 80-ft.[3] scuba cylinder confines more than a million pounds of kinetic energy. Equally important, bad air can form within. Gross corrosion in steel cylinders has been seen to reduce the oxygen content to below 12 percent, a deadly deficiency.

When poorly maintained cylinders fail to meet federal and industry standards, they must be condemned. Prudent preventive maintenance, on the other hand, will add many years of service life to a cylinder.

Corrosion

Corrosion is the major cause of damage to both steel and aluminum cylinders, because it means metal loss, which means structural weakness. Corrosion is easy for users to see on the exterior, but only if they examine the entire surface frequently. Users don't usually see inside their cylinders, so the assessment of the interior surface must be left to the judgment of whoever conducts the regular, formal inspections.

Unfortunately, some inspectors have no training or objective guidelines to follow. They simply don't know how to evaluate what they see inside and outside the cylinder. All technicians, whether in-house or at commercial facilities, should have formal, documented visual inspector training.

Water is commonly found inside scuba cylinders, where it can promote dramatic corrosion. Some divers believe that water enters their cylinder whenever the air pressure is low near the end of a dive, but this is untrue. Water most often enters the cylinder during the fill process. Poorly maintained compressor filter systems are probably the major culprit. Wet fills, in which a scuba cylinder is submerged in a tub of water during a fill, add their share as well. SCBA fill stations almost never use a water bath, and water in SCBA cylinders is decidedly uncommon.

Fill station operators who use water baths often cite their reasons for using them, but none of those reasons hold up to scrutiny. First, the water in most tubs is seldom changed, so the tanks aren't getting a freshwater washdown as professed. Second, water doesn't absorb explosive energy, so it provides no real protection in the event of cylinder failure. Rather, water transmits the explosive force to the container, which in turn will propel shrapnel throughout the work area. Finally, the heat generated by fast fills doesn't penetrate the walls of 3AL and composite cylinders fast enough for immersion in water to be of significant benefit.

Since water baths are inadequate, cylinders should be filled at from 300 to 600 psi per minute (by gauge) in a dry environment.

Also, as mentioned above, tank boots can be major promoters of corrosion, since they trap water against the exterior surface. Tank boots were originally made to allow round-bottomed steel cylinders to stand on end. Out of convention, divers often place them on flat-bottomed cylinders as well. On a flat-bottomed cylinder, however, a tank boot has no more worth than to protect swimming pool liners during training sessions. Your team would do well to dispense with them.

Cylinder Failure

Some scuba cylinders are no longer approved for service yet are still in use. Three types of aluminum cylinders manufactured under special permits SP6688, SP6576, and CTC890 must be removed from service. Hydrostatic retesters frequently return these illegal cylinders to service, as do poorly trained visual inspectors. The permits for these tanks expired years ago.

It is important not to confuse those expired permits with valid permits SP6498 or E6498 and SP7042 or E7042, which must be overstamped with "3 AL" by the hydrostatic retester. Even these acceptable aluminum cylinders must be correctly remarked before filling.

Physical damage is often overlooked by both the user and the fill station operator. A 3AL scuba cylinder in California was fitted with an unprotected stainless steel band. Over several months, contact between the dissimilar metals caused severe pitting to the cylinder walls. Several steel scuba cylinders with their boots left in place for long periods have either exploded or leaked through gross corrosion, conditions ignored by the owners, fill station operators, and inspectors.

Many vinyl-coated, ungalvanized steel scuba cylinders remain in service, despite their increased risk of explosive failure. Two have exploded in Florida. That type of cylinder warrants scrutiny at each formal inspection. Even minor tears in the vinyl cover allow water to enter and be held indefinitely against bare steel. Many fill station operators will ignore blatantly deformed vinyl caused by expanding corrosion, and will instead continue to fill those cylinders. Also, untrained visual inspectors may completely ignore external damage in vinyl-covered steel cylinders. Usually these cylinders are dated from the late 1960s through the mid-1970s. They are generally of yellow steel 70 ft.^{3}, 2,250 service pressure.

Who pays the price for explosive cylinder failures? About 90 percent of cylinder explosions occur during the fill process, meaning that the fill station operator is at the greatest risk. In addition to injuries—sometimes fatal, usually disabling—considerable property damage is a common result. Several dive stores have had cylinders rocket out through the roof. Other buildings have had their walls toppled.

How do those responsible for high-pressure cylinders ensure that nothing is wrong with them? Your team should demand that four entities be knowledgeable about the differences between good ones and bad ones. First, hydrostatic retesters must know their job and have current information about each cylinder sent to them. Their licenses (renewed every five years) and haz mat training (renewed every three years) must be current. They must conduct a visual tech inspection as well as hydrostatic tests. Second, the visual inspector who sees the cylinder at regular intervals between hydrostatic tests must also be trained and certified. The inspector must have objective, written standards by which to judge the cylinder. You're within your rights to ask to see those standards. Third, fill station operators must be trained, have current haz mat documentation, and know what external visual inspection is required before each fill. Ask your fill station operator for a copy of his checklist. Finally, the user should be able to identify potential damage and ensure that the cylinder promptly receives a technical examination.

Filling Cylinders

In the United States, employee FSOs must have federally mandated hazardous materials training within ninety days of hire, and then retraining every three years afterward (49 CFR 172.700). Federal or state OSHA regulations require that all cylin-

A properly trained inspector can determine whether a cylinder in use is still safe.

ders that pose a potential hazard to employees be safe as determined by visual inspection (29 CFR 1910.101(a)). Unfortunately, those regulations offer little information as to what a cylinder handler should know, and they provide nothing to guide an FSO.

Before air is pumped into a single cylinder, the air station manager should establish a safe working environment. The fill station itself should be made safe by incorporating whatever protective materials and procedures can reasonably be placed into the system. Although not all of the following will apply to every fill station, managers should consider these basic safeguards.

1. Place the fill station away from bystanders.
2. Consult with an engineer before constructing a cylinder-diversion device.

3. Configure the controls away from the cylinder receptacle.
4. Secure high-pressure hoses and fittings at two- to three-foot intervals.
5. Keep the fill station away from critical building supports and walls.
6. Provide a physical barrier, such as a concrete or steel shield, between the FSO and the cylinder during a fill.
7. Use an energy director, or explosion barrier, to send explosive force in a safe direction.
8. Regularly inspect the filters and piping of the compressor.
9. Allow only trained, authorized persons to work at the fill station.
10. Post operating procedures and safety alerts.

A representative of the branch of the DOT that oversees pressure vessels has admitted that, very likely, every dive store and many public safety teams with a fill station are in violation of some regulation, whether federal, state, or local. His recommendation was to "operate with caution and avoid accidents."

Every cylinder to be filled should first be subject to a visual examination. Some ill-informed owners might express concern and argue that no form of inspection is required as long as the cylinder's evidence of inspection (EOI) sticker has been applied within the past year. However, an EOI sticker only denotes that, on the day of the inspection, the cylinder met the standards of the technician who inspected it. If those standards are poor (too often they're nonexistent) or if the cylinder has been damaged in some way during the year following the inspection, the sticker might not be valid. There is no one-year guarantee, as some EOI stickers imply, and the valid period of an EOI sticker is only for a maximum of one year, contingent on use and abuse.

As a prefill inspection, the FSO should quickly check the cylinder for the following.

1. The presence of a DOT or ICC marking (CTC or TC in Canada).
2. A proper hydrostatic retester's mark within five years for 3A, 3AA, and 3AL cylinders.
3. An EOI sticker within the past year, or an equivalent record of inspection.
4. Evidence of heat damage, dents, bulges, or gouges.
5. Discoloration or other evidence of exposure to chemicals.
6. Corrosion penetrations of more than .015 inches.
7. The presence of a correct pressure relief device.
8. Aluminum cylinders marked SP6688, SP6576, and CTC890, which are illegal and should be retired.
9. Illegal round-bottomed aluminum scuba cylinders.
10. Steel cylinders should emit a clear bell tone when struck lightly.

In the dive industry, it's common practice to overfill cylinders for a variety of reasons, none of which are legally defensible. Some customers demand it, and some FSOs consider it customer service. Some FSOs will overfill a cylinder to ensure that it shows full pressure when the diver begins the dive. Still, a cylinder is full at indicated service pressure with an ambient temperature of 70°F. When the ambient temperature is higher or lower, the correct service pressure may also vary. If a cylinder is filled slowly in Miami in 100-degree weather, a 3,000-psig cylinder could read 3,180, since there is about a 6-psig increase for each degree Fahrenheit.

Divers preparing for an ice dive at 30°F might find their properly filled cylinders showing only about 2,760 psig. Rather than overfilling the cylinders, cold-water dive planners should take this drop in pressure into consideration. It's important to mention that most diver cylinder-pressure gauges and fill station gauges aren't accurate to within 100 psig. Also, the maximum operating temperature for gas cylinders in general is 130°F.

In several recent civil cases involving exploding cylinders, FSO qualifications and actions were severely questioned. Defense is difficult for FSOs who have no training and fail to conduct a prefill inspection. Most of the causes behind recent failures would have been seen by a trained FSO before charging the cylinder. Remember, the final responsibility for cylinder safety lies with the owner and the FSO. If you feel that your department or employer hasn't provided sufficient technical training with regard to cylinders, you should consider such training to be a priority.

Study Questions

1. Standards for scuba cylinders fall under what agency of the United States government?

2. The major cause of damage to both steel and aluminum cylinders is ________.

3. True or false: Most of the water that enters a cylinder does so when the air pressure is low, at the end of a dive.

4. Fill station operators often cite three principal reasons for immersing a cylinder in a water bath during the fill process. Name those reasons.

5. Since water baths are inadequate, how should cylinders be filled?

6. What was the original intended purpose of a tank boot?

7. As per the text, name three types of aluminum cylinders manufactured under special permits that must be removed from service.

8. In the United States, employee FSOs must have federally mandated hazardous materials training within how many days of hire, and then retraining how often afterward?

9. True or false: Any scuba cylinder bearing an EOI sticker issued within the preceding year may be considered safe.

10. A cylinder is full at indicated service pressure when the ambient temperature is ________.

11. For each degree Fahrenheit increase of ambient temperature, the pressure inside a scuba cylinder will increase approximately ________.

12. Given an ambient temperature of 30°F and a properly filled cylinder showing only about 2,760 psig, is it acceptable practice to add more air to bring the cylinder up to its 3,000 psig limit?

Chapter 8

The Initial Response

The importance of interagency planning cannot be denied. All responding agencies should be involved in developing joint response plans well before any incidents occur.

For dive calls, effective planning involves four essential steps. The first is to review past incidents so as to locate the potential problem areas in your district. Find out where people most commonly congregate around water, as well as where vehicles are likely to enter it. Note the potential flood zones. Besides learning the physical terrain of the district through the eyes of a public safety diver, it is equally important to assess the quality of responses in the past. Try to determine whether the command structure functioned effectively and whether or not any responders were put at unnecessary risk during the operation.

Once the potential sites have been identified and mapped, examine them for potential problems and hazards that they might present. Typically, sites are made difficult by steep embankments, obstructions of various kinds, and distance from the nearest road. If you're dealing with tidal water, you may be confronted with a varied set of possible circumstances for the same locale, depending on where the moon happens to be.

The third step involves determining what is required to mount operations in those areas, given the assets of the team and compounding problems such as weather and staffing shortfalls. Predictably, the final step centers on making all personnel and equipment ready to do the job.

Typically, dive sites are made difficult by steep embankments, obstacles, or various other kinds of impediments.

Rescue Versus Recovery

When the call comes in for a water-related emergency, one decision must be made right away: Will the response be for a possible rescue, or will it be a recovery operation? Typically, time is the determining factor. The rescue mode should be maintained for a given length of time after the incident occurred, usually from sixty to ninety minutes. Beyond that interval, it would be impossible to make a successful rescue of a drowning victim. Be sure that your SOGs specifically mention that your team will operate in a rescue mode for no more than a predetermined interval after a drowning incident. For the period after that, your SOGs should state that the team will stop to evaluate whether to continue operations in a recovery mode. If the weather has deteriorated or if some other circumstances have intervened, it may be best to shut down the operation, perhaps to start up again when conditions improve. The recovery of a corpse is never worth the life of a responder.

Gathering Information

Suppose a mother calls 911, saying, "My child and his friend were playing by the lake, and now I can't find them!" The dive team deploys and searches the indicated area, but its members turn up nothing. The hours drag by, personnel push themselves to their limits, money is consumed, and hope grows dim.

Around six o'clock, somebody at home answers the telephone. The child is calling from his friend's house to find out whether he can stay for dinner.

What happened in this scenario is not entirely unusual. During the rush to deploy the team, no one thought to determine whether the child might have gone elsewhere. Numerous teams have committed themselves to operations only to find out hours later that the person they were searching for was safe all along.

While you are interviewing witnesses and the dive team is setting up, assign a team member to do some research. If no one actually saw the person go under, find out where he or she might have gone instead. Too many would-be victims are later discovered at home, on the street, with friends, or in bars for you to ignore this rather demoralizing possibility.

Interviewing Witnesses

The first step in interviewing witnesses is to pick up two critical tools: paper and a writing utensil. Without writing the information down, you won't remember exactly what was said, you won't be able to relay it accurately to someone else, and you won't know the second time you interview that same witness whether you've received consistent information or not. Without written notes, you won't be able to correlate information from the scene, different witnesses, or different interviews of a single person. Written records allow you to review the incident later, and written statements from witnesses can be vitally important if the incident turns out to involve foul play. Of course, having information written down helps cover you and your team in the event of a lawsuit. The rule holds firm: If it isn't on paper, it didn't happen.

Try to put yourself in a physical location where you can see as many of the witnesses as possible, as well as the water and the surrounding land. Doing so will help you maintain a position of control. It will also help your inner mind put together the pieces of the incident. When you are visually in control of what is happening, your inner mind will begin to decipher the circumstances and the possible location of the target. If you are in a position of low control and constantly need to revisualize what has taken place, you aren't likely to make such connections. In short, never turn your back on the water, if at all possible.

At the beginning of a call, only one person should conduct the interviews. This will help a witness develop trust with his interviewer and lessen the feeling that he's being interrogated. Whenever possible, interviews should be conducted by law enforcement personnel, since their training and duties give them experience in the thorough questioning of witnesses. Otherwise, you should assign a capable member of the team who is trained in conducting interviews.

As you record the statements of witnesses, be sure to include every detail that they provide. If a witness states that he saw a person wearing a red hat, you must write down that the person was wearing a red hat. Although the color of a person's clothing may seem insignificant at the time, it could be very important, since color and shape can be clues as to how far the witness was from the victim. Too often, key words aren't recorded because they aren't recognized as key words.

A thorough interview should consist of three parts. The first part is "Show me," in which the witness physically reenacts everything that he did just prior to and during the incident. The second part is "Tell me," in which he describes what he saw. The third part is "Write it," in which the witness commits his statement to paper, plus any drawings that he feels are necessary.

The "Show me" stage is essential. Being witness to a drowning can be emotionally traumatizing, and the human mind will often block out conscious memories of the event, especially if feelings of guilt or helplessness are involved. Although the mind may short-circuit memories, the body won't forget. Body language during a reenactment becomes invaluable. Having a witness reenact his actions will stimulate his brain, activate memory paths, and possibly bring forth information that he might not have remembered otherwise. If a witness says that he ran to the edge of the dock, have him run to the edge of the dock. If he says that he called for help, have him do so in the reenactment.

The body may also be more truthful. If a witness tells you that he ran to the water's edge, and now he's hesitant to go out to the same edge, perhaps he never went there. It's not at all uncommon for a person's body language to tell a story that's different from the verbal description he offers.

Witnesses can also reenact what someone else did. If, for example, he tells you that someone threw a small object into the water, have him demonstrate it for you. The type of throw will almost certainly be imitative, and it will give you a clue as to how far from shore you need to search.

In reading the body language of a witness, closely observe his overall stance. Suppose that a witness points out over the water, stating what he saw. His arm goes out straight, but his torso and legs stand at a different angle to the shore. What he may be indicating is that, because of the trauma involved, only his subconscious mind has a clear idea of what actually took place. In observing the incident, it's almost certain that the witness faced it square-on. Thus, the subconscious mind will often orient a witness toward the proper location, even when his eyes, pointing finger, and best intentions do not.

In a process called triflexing, the interviewer should take such a witness to three separate locations that look out onto the scene. One of those locations should be where he reportedly stood at the time. At each place, the witness should point to the spot where the incident occurred, as he remembers it. The interviewer should take

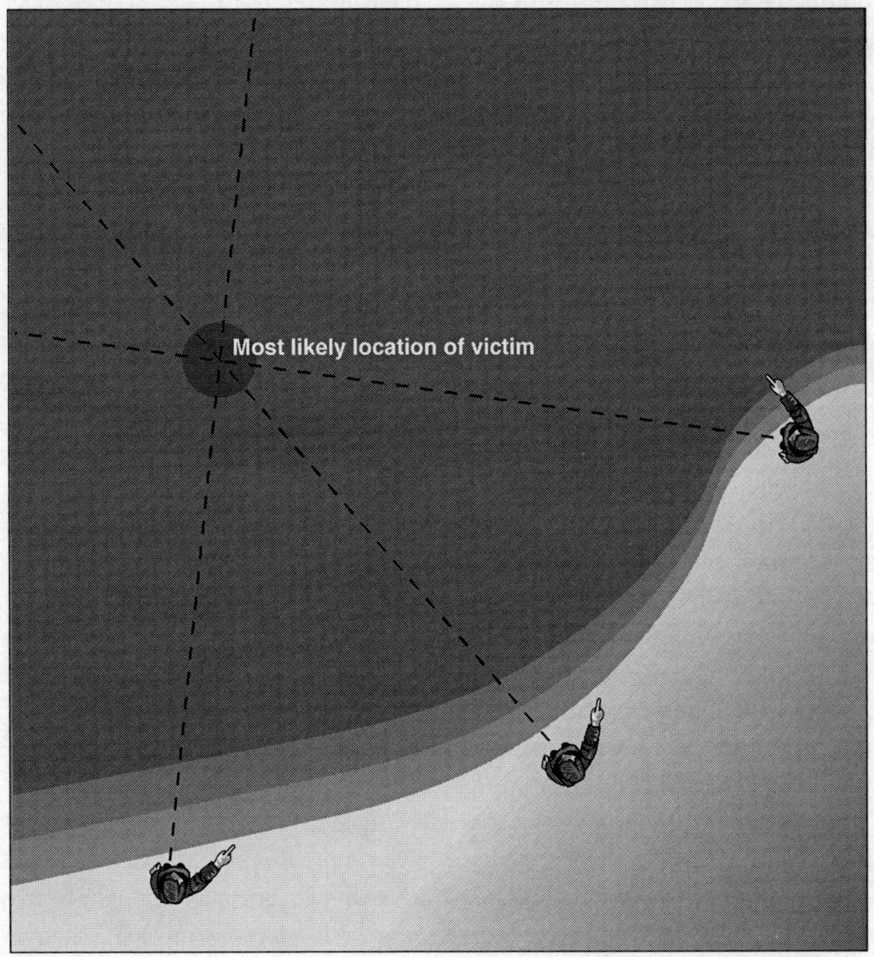

Triflexing focuses on the orientation of a witness's body as a way of determining the most probable location of the victim.

note of the position of his body as he does so. Later, on a profile map, the interviewer should draw lines indicating which way his body was facing versus which way he was pointing. Often the lines drawn from his torso will converge, and that point of intersection is the most likely location of the person or object in question. If a witness's body doesn't face straight-on to the target point, it's usually because either the investigating officer programmed him that way or because he's lying. In actual practice, the triflexing system often gives better information as to location than the witness's verbal statement.

After the reenactment, ask the witness to close his eyes and picture the event in his mind's eye. Ask whether he noticed anything on the opposite shore while the drowning occurred. Remember that, in focusing on the victim during a moment of crisis, it's entirely possible that the witness saw nothing else. Thus, you shouldn't be too persistent in this line of questioning, lest the answer you receive become one of invention.

The majority of drownings are witnessed, and the information that witnesses provide is vital. Still, the average witness will underestimate distances over water by at least 25 percent. This underestimation can be addressed two ways, and both work best when they're used together. The first method is by way of simple arithmetic: Add 25 percent to the witness's estimate. If a witness says the victim drowned forty feet from shore, add ten feet to the guess, and start looking at fifty. The second method involves sending out a diver along the surface until the witness tells you he's at the correct spot. Since the diver is on a tether, you'll have a direct indication of distance. If you can't send out a diver for some reason, toss an object into the water and have the witness estimate the right location in relation to where it hits.

The statements of witnesses have to make sense. Don't simply accept what a witness tells you—ask yourself whether it's logical. If the information you get doesn't seem right, the witness could be lying or he may have facts distorted in his mind. Sometimes a person who's prone to being overhelpful will exaggerate or even make up stories to make his role in the incident seem more important. Bear in mind that children often make excellent witnesses, because they typically won't lie or create information as adults might. One unfortunate aspect of interviewing them is that, if adult witnesses are available, children are often overlooked. Pets, too, have been known to indicate with a high degree of accuracy where their owner went down. Don't limit your witness options. Try to keep a global view of the scene.

You should interview witnesses separately, since they may otherwise end up elaborating, collaborating, and collating their stories, thus altering the facts. As you conduct an interview, avoid leading questions, since they require scant thought to answer and usually evoke affirmative responses. Instead, ask questions in such a manner as to draw out information. Instead of, "Were the victims near that large rock?" ask, "Where in relation to that large rock were the victims?" Try to let the witness remem-

If the information you get doesn't seem right, the witness could be lying or may have unintentionally distorted the facts.

ber what was taking place rather than prompting him with suggestions. In the same vein, avoid touching a witness, even to help him to a location. Let him guide you. If a witness can't get to a location now, ask yourself how he might have gotten there at the time of the incident.

An interviewer needs to be a good listener. Demonstrate that you're interested in what a witness has to say. Be patient, positive, civil, and nonjudgmental. Try to understand people and human nature. Relate to the witness without becoming involved. Drowning is an emotional event for the family, witnesses, and responders, so it may help to take a moment or two to stand back and try to understand better what it is they're feeling. You should always take care of witnesses. Remember that a witness may have been in the water during the event and that he could be suffering from shock or cold stress. Before taking him back to the site, make sure someone dries him off and gives him something warm to put on. If the reenactment involves going back into the water, make sure that he puts on a PFD so that he won't feel uncomfortable being there again.

Whenever possible, avoid secondhand information. Only an actual witness can tell you what happened. Still, friends and family members who didn't witness the incident may be able to offer useful information as to the habits of the missing individual. To see whether the statements of witnesses are logical, it may be appropriate to ask nonwitnesses such questions as "Where did your friend normally fish on the lake?"; "Have you ever seen your child playing alone by the water?"; or "With whom does your child normally play after school?"

Following is a list of basic questions to ask witnesses.

1. What alerted you to the incident?
2. Did the victim call for help?
3. Where were you when you first saw or heard the incident?
4. What were you doing when you first saw or heard the incident?
5. Can you please reenact exactly what you did?
6. Can you please describe and demonstrate the actions of the victim?
7. Were you alone?
8. Were the victims alone?
9. If the victim wasn't alone, where were the other victims?
10. How long ago did this happen?
11. What color clothing were the victims wearing?
12. In what direction were they moving?

Not All Drownings Are Accidents

Public safety divers and other team personnel must always consider the possibility that a person didn't end up dead at the bottom by accident. In some cases, the evidence of foul play will be clear. Not every homicide victim drawn up from the depths has bullet holes in his head and cinderblocks tied to his ankles, however. Some perpetrators will try to cover up a murder by making it appear as if it were an accident instead, and these are the situations of which team members must be aware. The topic of intentional drownings is too large to discuss properly within the scope of this book, and the investigation of such crimes should be handled only by law enforcement personnel. Still, anytime a member of a dive team believes that a witness or family member is acting suspiciously, he should report it to the incident commander, who should then ask a law enforcement officer to interview that person as soon as possible, though without raising suspicion. Dive team personnel should in particular consider witnesses or family members who change their stories, add too much detail (as though they're trying to convince you that they're telling the truth), spontaneously offer that they "had nothing to do with it," want to leave as quickly as possible, or who don't seem to care about the victim. Team personnel should also be suspicious if the victim was wealthy, shows signs of physical abuse, or was handicapped.

Determining Where to Search

It isn't enough to know where a victim went under. The question is, where is he now?

Wind and Currents

Although wind won't affect a body or any other object on its way to the bottom, it most certainly can have an effect on a victim struggling on the surface or something that takes awhile to sink, such as a car or a boat. The wind will move an object in direct proportion to both its strength and the surface area of the object above the water. Consider a person drowning at the surface, with only the head and shoulders exposed to the wind. Since the head and shoulders account for approximately 10 percent of the body's surface area, this person will be affected by 10 percent of the wind. If the wind is blowing at 2 knots, or approximately 200 feet per minute, the person in the water will be pushed along by the wind at 20 feet per minute (200 ft./min. x 1/10 = 20 ft./min.). If a person takes one and a half minutes to submerge, the wind will therefore push him 30 feet before he goes under (20 ft./min. x 1.5 min. = 30 ft.). (These values are only estimates and do not account for drag in the water.)

Obviously, water currents will also move an object. At the surface, the object will be carried at the same speed and in the same direction as the current. Thus, if the current is flowing at 1.2 knots, or about 120 ft./min., and if the object takes a minute and a half to go under, it will travel 180 feet before dropping out of sight (120 ft./min. x 1.5 min. = 180 ft.).

If the wind and the water in the two previous examples are flowing together, then locating the submergence point is a matter of adding the distances together (30 ft. by wind + 180 ft. by water = 210 ft.). If the current and wind are moving in different directions, you can calculate the position by trigonometry. A calculator with a trigonometric function may prove useful in this regard. Generally, the influence of winds of less than about thirty knots can be estimated, since the current will have the greatest effect.

Sink Rates

On average, a human body will sink at a rate of about 1.5 ft./sec. in salt water and at 2 to 2.5 ft./sec. in freshwater. Variations in muscle-to-fat ratios, body composition, and clothing will make a given body sink slower or faster than the average. Because of these variables, the best rule of thumb is to estimate that bodies will sink at a rate of 2 ft./sec. in any environment. Thus, a body should take about 20 seconds to reach the bottom in water 40 feet deep.

Except in extremely shallow water or in the presence of extremely fast currents, once a body lands on the bottom, it will not move—even in a current—until it has putrefied enough to gain buoyancy from the gases produced within. Once the body has enough buoyancy, it may only float partially. It may drift and scrape along the bottom for some time before finally rising to the surface.

Dive teams often state that, once a body reaches the bottom, it will travel hundreds of feet, or even many miles, but this is untrue. The Buffalo Police Department Underwater Recovery Team, for example, reports that in the Niagara River, which can move at 10 to 15 knots (1,013 to 1,520 ft./min.), bodies remain in one place on the bottom until they start to float.

One of the reasons that dive teams may think a body doesn't stay in one place on the bottom is because they don't find it until it has already reached the surface. By that time, of course, it could easily be miles from the descent point. It is likewise a myth that a body suddenly becomes buoyant and pops to the surface directly over the spot where it was lying on the bottom. Rather, it may actually travel a great distance underwater until it gains enough buoyancy to overcome the temperature strata of the water and its own wet weight.

If a body sinks in water in which there is little or no current, finding it should be relatively easy. Normally, it will be on the bottom within an imaginary circle, the radius of which is equal to the depth. If a person drowned in 32 feet of still water, his body should be somewhere on the bottom within 32 feet of the point directly below the point of submergence.

If a body descends in a current, of course, the forces acting on it will be more deliberate. To find the speed at which a given current is moving, there are two practical ways to estimate it. The first method is to toss in a tennis ball tied to fishing line that has been knotted at five-foot intervals. Determine how long it takes the ball to travel 100 feet. (You can track the ball for 50 feet, if that's more convenient, then double the result.) Since one knot equals a rate of approximately 100 ft./min. (101.333 for purists), you can determine the speed in knots by dividing the rate constant 1 (for one knot) by time. Thus, if it takes 48 seconds, or .80 minutes, for the ball to travel 100 feet, 1 ÷ .80 = 1.25 kts. The second method is to see how far the tennis ball will travel in one minute. Divide the distance by 100 (or, again, 101.333) to find the speed in knots. Thus, if the ball floats 220 feet in one minute, 220 ÷ 100 = 2.2 kts.

In point of fact, calculating the current in knots is an elementary but wasted step, since what we're normally interested in for everyday, small-scale calculations is the ft./min. figure.

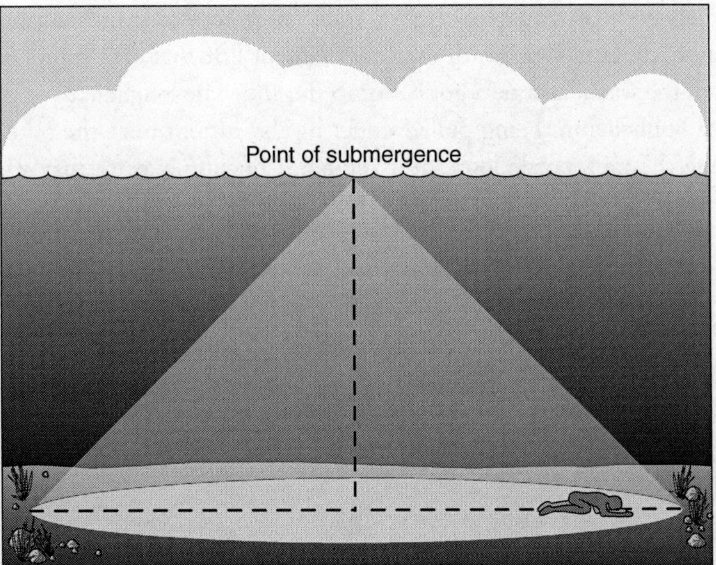

If a body sinks in still water, it will normally touch bottom within an imaginary circle, the radius of which is equal to the depth.

The formula for determining where a body is likely to touch bottom is as follows

$$\frac{\text{Current in ft./min}}{\left(\frac{\text{60 sec.}}{\text{Time to hit bottom in seconds}}\right)} = \text{Downstream drift underwater}$$

Given a surface current of 200 ft./min. and a 40-ft. depth, the solution works out as follows.

$$\frac{\text{200 ft./min.}}{\left(\frac{\text{60 sec.}}{\text{20 sec.}}\right)} = x$$

$$\frac{\text{200 ft./min.}}{3} = x$$

$$\text{66.666 ft.} = x$$

From these calculations, this body traveled nearly 67 feet downstream from the point of submergence before touching bottom.

When all else fails, you can weight a milk bottle so that it sinks at the same rate as a human body. Attach a length of monofilament line that's five times longer than the depth of the water, and tie a bobber onto the end. The long length of the line will prevent the bobber from being pulled under by the current once the bottle hits bottom. Setting the milk bottle loose near the point of submergence may well put you near your target.

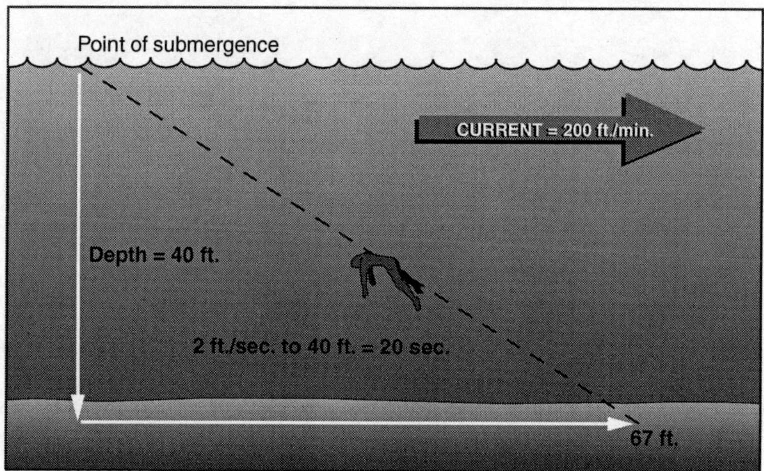

The location of a body can be estimated based on the point of submergence, the strength of the current, and the depth of the water.

Grids

At best, the results of such calculations are only estimates, and they presuppose that the point of submergence is known. It's likely that the search area will be fairly extensive. The first step in narrowing the search is to eliminate those areas of impossibility or lesser likelihood. On a large-scale map, whether hand-drawn or otherwise, divide the body of water to be searched into a grid. Based on witness interviews and the observed currents, mark out those areas upstream of the submergence point and those that are clearly too far or too near. Once you've outlined the overall search zone on the map, you can place outer-perimeter buoys in the water according to the grid. Place these buoys 50 to 75 feet apart to form an L or a complete box, designating the entire search area. The divers will search in patterns within this boundary, using the buoys and possibly ranges from shore as guidelines. The actual distance between the outer-perimeter buoys is determined by depth. If the water is deep, the less area around each buoy that the divers can cover; thus, the buoys should be placed closer

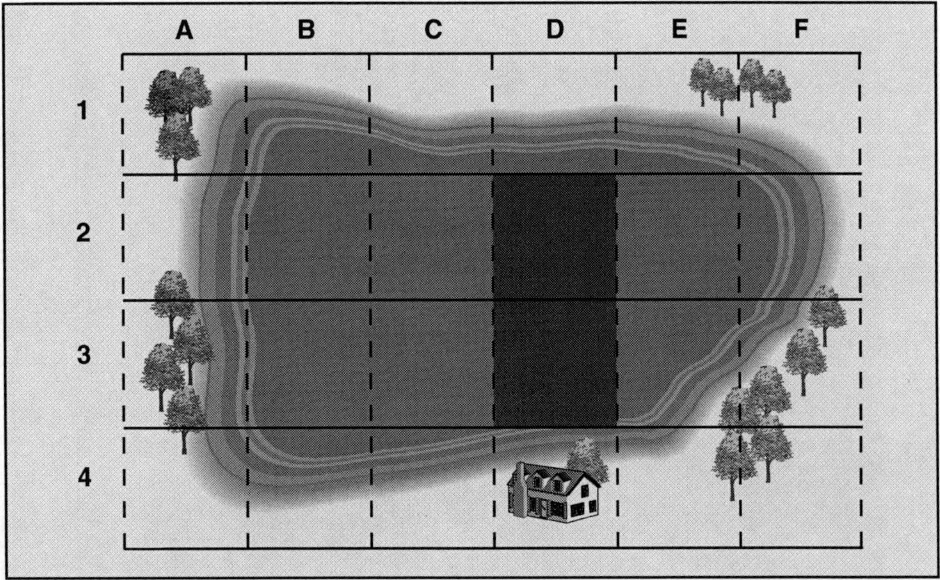

The first step in narrowing a search is to eliminate those areas of lesser probability, based on known landmarks and witness observations.

together. When a diver completes a search, thereby securing an area, additional buoys can be dropped within the box to section off each area. It's better to do this after rather than before the search to prevent the diver from becoming snagged in the buoy lines.

In determining distance, color can be helpful, since colors are more difficult to distinguish from far away. Other details can be useful as well. If a witness on the shore was able to make out the name on the transom of a boat, then clearly the incident couldn't have taken place beyond reasonable limits. Even being able to distinguish gender may be a clue. In cases where such details are offered, the investigators must consider the lighting as well as the eyesight and perspective of the witness to estimate how far out an incident may have occurred.

Ranges

One of the most important tools to use when interviewing witnesses and working away from shore is a range. A range can help you document search areas on profile maps, as well as your own location aboard a boat. By taking a range, you'll be able to know whether the boat you're on has moved or not. Ranges can be particularly helpful if you're working in an area where the water's edge may vary, as in a tidal zone.

To determine a range, choose a target object on the distant shore. Find an object behind it that lines up with it directly. Next, look to your right or left, and find two more objects that line up in the same manner. After leaving that spot, you should be

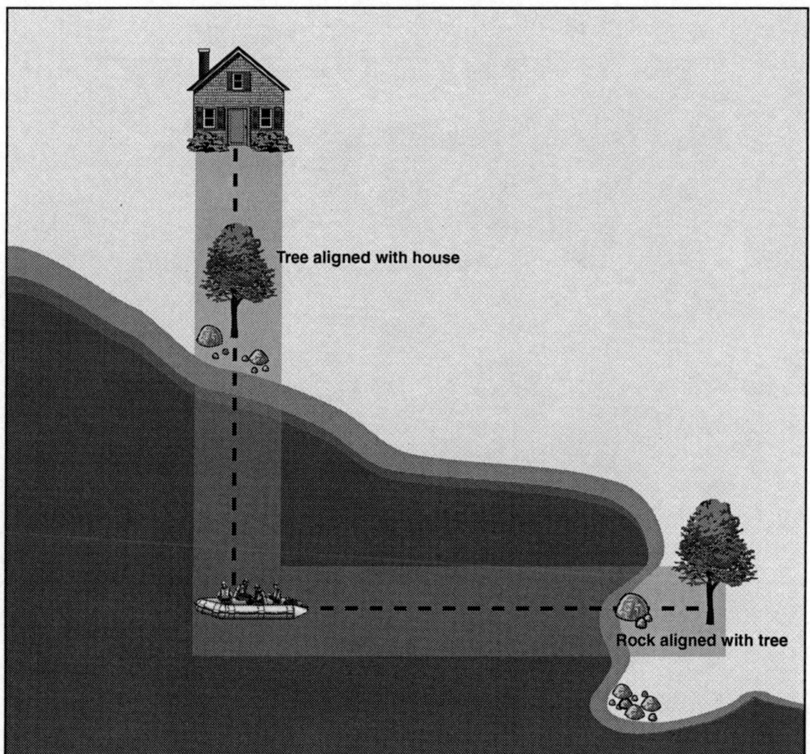

By taking a range, you can document your position aboard a boat.

able to come back to it again by aligning those four objects as they were before. Having the same vantage of them puts you back at the original location. You can record compass readings or even take Polaroid pictures for accurate documentation of the ranges.

Risk-to-Benefit Analysis

The decision as to whether to make the dive at all rests on a risk-to-benefit analysis conducted by the dive coordinator. It is this officer's task to assess the conditions at the incident scene, including the readiness and capability of the responding units, and therefrom to decide whether or not to proceed with a diving operation.

The preparedness of responding personnel is as important as any matter of environment, time of day, contamination, or extenuating hazards. The average public safety diver has made fewer than thirty dives. The average team captain has made fewer than seventy-five dives, half of them recreational. Public safety divers, who will be called to descend into black, debris-filled water at all times of the day and night, should be considered inexperienced if they have made fewer than fifty dives, unless the majority of those descents were made in the course of public safety missions. A diver who has made most of his three hundred descents in water shallower than forty feet may be less prepared to go to a hundred feet than a diver who has been there sixty times but with far fewer overall descents. Just because someone is a strong, experienced black-water search diver at depths of thirty feet or so doesn't mean that he has the skills and equipment to make a safe dive to ninety feet, even with good visibility.

Air consumption increases with depth. Compared with breathing at the surface, a diver will breathe twice as much air at thirty-three feet, three times as much at sixty-six feet, and four times as much at ninety-nine feet. Typically, a public safety diver searching with an 80-ft.^{3} tank has a surface air consumption rate of 20 to 65 psi/min. Using 50 psi/min. as a conservative average, he will use 100 psi/min. at thirty-three feet, meaning 2,000 psi in twenty minutes, and 150 psi/min. at sixty-six feet, or 3,000 psi in twenty minutes. A heavily stressed diver will use even more air, thus pointing out the severe limitations of a 19-ft.^{3} pony bottle. Air depletion is one of the most common causes of underwater injuries and fatalities. A method of calculating a diver's surface air consumption rate, and from that, to derive his depth air consumption rate, appears in the Appendix.

Nitrogen narcosis can start anywhere from seventy to a hundred and thirty feet, although most divers don't recognize its early symptoms. High carbon dioxide levels, cold, fatigue, and stress can all increase the risk and severity of this effect. Nitrogen narcosis can cause tunnel vision, single-mindedness, risk-taking behavior, and paranoia. Given the influence of nitrogen narcosis on the thinking process, a public safety diver working under difficult circumstances may panic and bolt to the surface as a result of something he encounters on the bottom, whether truly problematic or not.

If a diver panics for whatever reason, he will most likely try to reach the surface as quickly as possible. Too many divers who make such uncontrolled ascents will rip off their mask and spit out their regulator on the way up. If a diver is out of air because his tank his empty, because he has shunned his regulator, or because choking has resulted in laryngospasm, he may well drown if it takes him more than a few seconds to reach the surface. If the diver is choking on aspirated water, a common cause of panic, he may hold his breath during the ascent. Doing so is a risk factor for lung overexpansion injury.

By definition, a backup diver is someone who can reach a primary diver within sixty to ninety seconds of a signal for help, but even this short span of time can be an eternity in an emergency situation. Consider the backup diver who passes off a pony bottle to an entangled primary diver, then returns to the surface to replace the pony and obtain a contingency tank of 80 ft.^{3}. Given the high consumption rates at depth, a pony bottle will have a useful life of only a few minutes during a crisis. At forty feet, such an emergency would be manageable. At a hundred feet, the primary would be in trouble.

According to Henry's Law, increased pressure results in an increased solubility of a gas into a fluid. The deeper a diver goes, the more nitrogen is dissolved in his tissues. If he ascends too quickly, he will experience a form of decompression sickness commonly known as the bends, resulting from the release of nitrogen gas back into his system. With a high pressure load, slow ascent rates become critical, as do long intervals between dives. Making a safety stop, or pausing at a depth of about fifteen feet for three to five minutes, will help to decrease the risk of lung overexpansion injury. Even more importantly, it will help to decrease the risk of decompression sickness, especially for dives in excess of sixty feet. Still, very few dive teams that go deeper than sixty feet bother to use safety stops during the ascent phase.

Suppose that a public safety diver becomes entangled at the end of a twenty-minute dive to one hundred feet. If it takes the backup diver six minutes to reach and extricate him, the primary diver will already be beyond the no-decompression limit by anywhere from one to eleven minutes, depending on whose dive tables the team is using. This diver must make a mandatory decompression stop during his ascent, yet he may not even be aware of that need if the water is too murky to read gauges and tables, or if underwater communications are lacking. Many such problems face dive teams, and the risks of decompression sickness abound. Fewer than half of all public safety divers, and far fewer tenders, are proficient with no-decompression dive tables, and few divers can make a three-minute safety stop at fifteen feet in black water without holding on to an ascent line. Some teams equip their divers with lift-capable body bags for recovery operations—the diver can simply send the corpse to the surface rather than having to hold on to it during a slow, controlled ascent—but, unfortunately, this is far from standard practice.

During extended operations, rotating divers back into the water after making a deep dive can be problematic or impossible. Given a small operation of three divers and two tenders, with maximum dive times of twenty-five minutes to a depth of fifty feet, each diver can make three descents with about a fifty-minute interval at the surface. Within these parameters, dive tables, decompression sickness, and mandatory safety stops become less of an issue. For dives below fifty feet, however, surface intervals need to be extended, and safety would soon be compromised if only three divers were in rotation.

Considering all of these factors, the maximum safe depth for most relatively inexperienced public safety teams is fifty feet, with a possible ten-foot extension if approved by the dive coordinator. The decision to go to sixty feet would depend on the known air-consumption rate of the diver, his current rate of consumption, the water conditions, and the general circumstances of the incident. Dives to between sixty and one hundred feet require 30-ft.^{3} quick-release pony bottles, plus greater training and experience. Limited-visibility search dives below one hundred feet should only be conducted with surface-supplied air, bailout bottles, communication systems, full-face masks, and extensive training.

In making his decision, the dive coordinator must weigh the risks to the team against the potential benefits of the operation. If a child falls through the ice on a lake about thirty feet deep, without apparent trauma, there is a reasonable chance that divers might save him if they can recover him within half an hour or so, since the maximum known survival time for a cold-water drowning is sixty-six minutes submerged. The risks to a responding dive team in such a scenario are rather small. There's always a chance that a diver could become injured or hypothermic, but weighing the potential benefits, the dive coordinator would likely be correct in deploying his team for a rescue operation. Other circumstances, such as an overhead environment or downed power lines, might easily countermand this decision. If the victim was in his forties and the submergence took place in proximity to the boil of a dam, for example, the chances of a rescue would quickly approach zero, and the attendant risks might make even a recovery ill-advised.

In making his decision, the dive coordinator must weigh the risks to the team against the potential benefits of the operation.

Study Questions

1. What are the four essential steps in planning for a response capability?

2. The rescue mode should be maintained for a given length of time after the incident occurred, usually from ________ to ________ minutes.

3. At the beginning of a call, why should only one person conduct witness interviews?

4. A thorough interview consists of three parts. What are they?

5. Why may body language be more accurate than the verbal description a witness gives?

6. What is triflexing?

7. According to the text, the average witness will underestimate distances over water by at least ________ percent.

8. Why can children make excellent witnesses?

9. One knot equals approximately how many feet per minute?

10. On average, a human body will sink at a rate of approximately ________ feet per second.

11. If a person drowned in fifty feet of still water, his body should be somewhere on the bottom within ________ of the point directly below the point of submergence.

12. When staking out the perimeter of a search grid, the distance between the buoys is determined by ________.

13. What is a range?

14. The decision as to whether to make the dive at all depends on a ________ analysis performed by the dive coordinator?

15. Compared with breathing at the surface, a diver will consume ________ as much air at a depth of thirty-three feet, and _______ as much at sixty-six feet.

16. What is Henry's law?

17. What is a safety stop?

Chapter 9

Deploying the Team

Many public safety divers, especially inexperienced ones, become physically fatigued or even overwhelmed simply by putting on their gear. Given the urgency and expectations of an emergency call, it isn't uncommon for divers to become so exhausted from arriving and dressing that they cannot dive safely for the operation. This exhaustion may be due to any or all of four causes: poor mental preparation, poor physical fitness, poor procedures, and being overloaded with tasks. In addressing three of these problems, divers must remember that their sole function at an incident site is to act as the eyes of their tender. They need not be concerned with witnesses or other extraneous matters; rather, they need only follow the procedures as they have been trained. If those procedures have been well thought-out, they will be the simplest, most organized way to perform any given task. A diver should never rush, because that will only increase the chance for error. In fact, it may do him well to pause for thirty seconds or so before entering the water, just to clear his mind and focus on the mission before him. If the cause of his exhaustion is poor physical fitness, then the diver needs to address the problem by other means—either that, or be content to serve in a support capacity. Out-of-shape personnel don't belong underwater, tethered or not.

Predive Fitness Check

Before a diver is asked to suit up, his tender should ask him two questions: First, has he consumed alcohol within the past twelve hours? and second, does he feel capable of diving now? If the diver has been drinking, has difficulty equalizing, has stuffy sinuses, is already cold, or exhibits other contraindications to diving, he should be asked to assist with other duties. If you as a diver find yourself saying, "I'm sure I'll be fine once I get in the water," then you don't belong in the water. Recent alcohol consumption by anyone on the team may exclude him from performing in any capacity, both for safety and liability reasons.

The tender should carefully observe the diver. If the diver displays any reluctance to dress or dive, this should be reported to the dive coordinator. Out-of-character behavior should be noted and questioned. The tender's main responsibility is always for the safety and well-being of the diver.

As part of the predive check, divers should have their blood pressures taken before suiting up. Ideally, all divers and tenders should learn how to take blood pressures. If the diastolic pressure is over a maximum limit, such as 100 mmHg, then the diver should not be allowed to dive until his pressure has dropped.

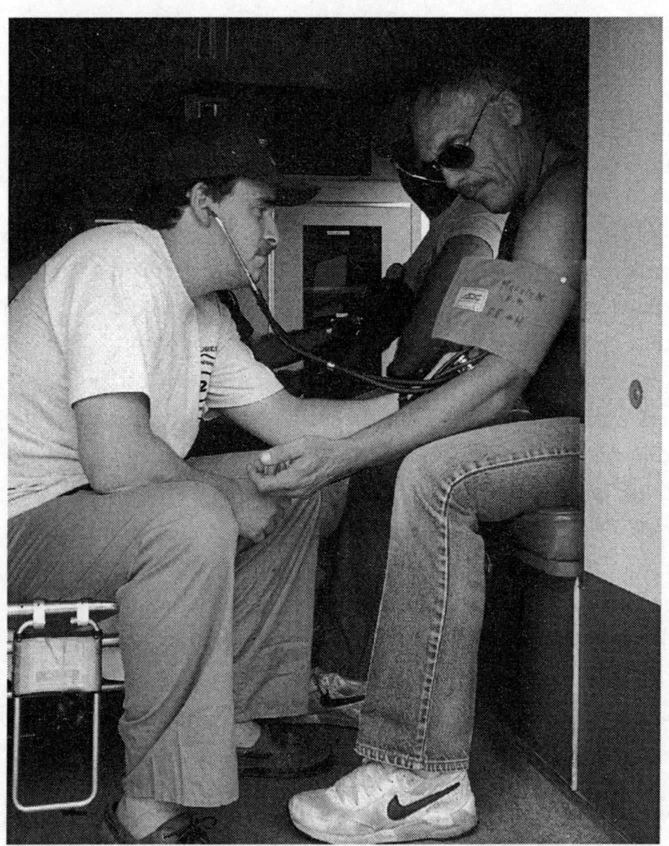

All divers should have their blood pressure checked before being allowed to suit up.

There are two ways of deciding whether the systolic pressure is too high. One way is to add the diver's age to 110 mmHg if the diver is a male or to 100 mmHg if the diver is a female. If the systolic pressure is higher than that sum, then the diver should not be allowed to go underwater. Otherwise, the department can set a maximum limit, such as 160 mmHg, as the determinant. The criteria should be explicitly defined in the department's SOGs.

If an operation is in the rescue mode and it isn't possible to check a diver before he enters the water, then EMS personnel should thoroughly check him afterward. Each diver should also consult his personal physician regarding blood pressure and underwater operations.

Gearing Up

Some teams consider themselves recovery teams only, and so never practice for a rapid deployment capability. This is a mistake. The deliberate, professional manner in which an experienced rescue team deploys should be practiced by anyone who operates as an agent of public safety. The very nature of public safety means that per-

Public safety personnel must be ready to deal with a variety of unexpected scenarios. (*Photo courtesy of Houston Police Dept.*)

sonnel must be ready to deal with an endless variety of unexpected scenarios. Even a recovery team that is never dispatched until after rescue has been deemed impossible must be ready for that unforeseen day during a declared state of emergency when a school bus becomes trapped in the middle of a flooded intersection and other responding units are delayed by the aftereffects of a storm.

Wet Suit

Wet suits can either be powdered for quicker donning, or they can be primed with liquid soap. Otherwise, divers can purchase specially lubricated versions that can be pulled on easily and with reduced wear, even when the suit is wet.

To don the lower part of a wet suit, extend one leg into the suit until it stops, then bring the suit over your foot and ankle. Work out any wrinkles up to your knee before you insert the other leg. As you don the suit, be sure to work out any wrinkles. Otherwise, when you descend, the air pocket within that wrinkle will contract, pulling your skin into it. After you've experienced one of these painful pinches, you'll never forget to work out the wrinkles again.

If the suit jacket has short legs with a zipper down one leg, close the zipper four to five inches, then step into the jacket as though you were putting on a pair of shorts. Closing the zipper before donning the jacket is easier than trying to close the two halves around your thigh. Once the shorts are on, put on the rest of the jacket. If you have a suit with a beaver tail, start with the jacket first. As with the legs, insert an arm until it stops, then work the cuff around the wrist. To achieve the best fit, insert one arm until it meets resistance, then extend your second arm into the suit and stretch forward with both arms. Work out all of the wrinkles up to the shoulder area without any excessive pulling or tugging. Your tender should assist you as necessary to help prevent fatigue.

Don the booties as you would socks. Side-zipper booties are preferred, but if your booties have no zippers, cut small slits at the toes to allow trapped air to escape as your foot works its way in. Wear your booties under the cuff of your wet-suit legs so that they won't fill with water.

Put on your hood before zipping the collar of your wet suit. To don the hood, roll the skirt in your hands, then pull the hood over your forehead, with your head bent slightly forward. Pulling the hood over your head rather than your forehead may compress your cervical spine, possibly causing pain or injury. After the hood is on, zip the suit completely.

Dry Suit

Before donning a dry suit, put on some type of undergarment. Even if you only wear sweats, you need something to keep your skin from getting pinched by any wrinkles in the suit as you make your descent. Also, gently exercise all of the seals in the suit, especially in cold weather. Stretch and release them slightly about ten times to warm them up, to make the rubber more pliable, and to reduce the chance of tearing.

Donning the suit itself is a simple task. Step into the legs first, then bring up the suit to your waist. Do not tug or pull to excess. As with a wet suit, you should slowly work the wrinkles upward. The suit should go on smoothly. To put on the wrist seals, hold your hands straight with your fingers together. As you push your hands through, avoid puncturing the seals with your fingernails. If you need to pull the seals up farther onto your wrist, don't pull the seal itself. Rather, pull on the material above where the seal attaches. Wrist seals should sit above the bony protuberance of the wrist. Seat them flat by gently reaching in with your fingers and pulling the seal into place.

Next, grasp the left and right sides of the neck seal, and gently pull the seal apart as you roll your forehead through it. Keeping your hands apart will reduce strain on both the suit and your neck. As with the wrist seals, gently use your fingers to seat this seal properly.

The zipper is the key component to maintaining a dry system. Avoid twists, bends, and snags. The tender should close the zipper on a dry suit, even if the suit is of a front-entry style. The diver will be pumped up and ready to get into the water, and he may not take the time to close the zipper properly or ensure that no clothing is in the way. For a back zipper, the diver should stand with his arms raised forward to shoulder height, and he should be bent forward to flatten the zipper as much as

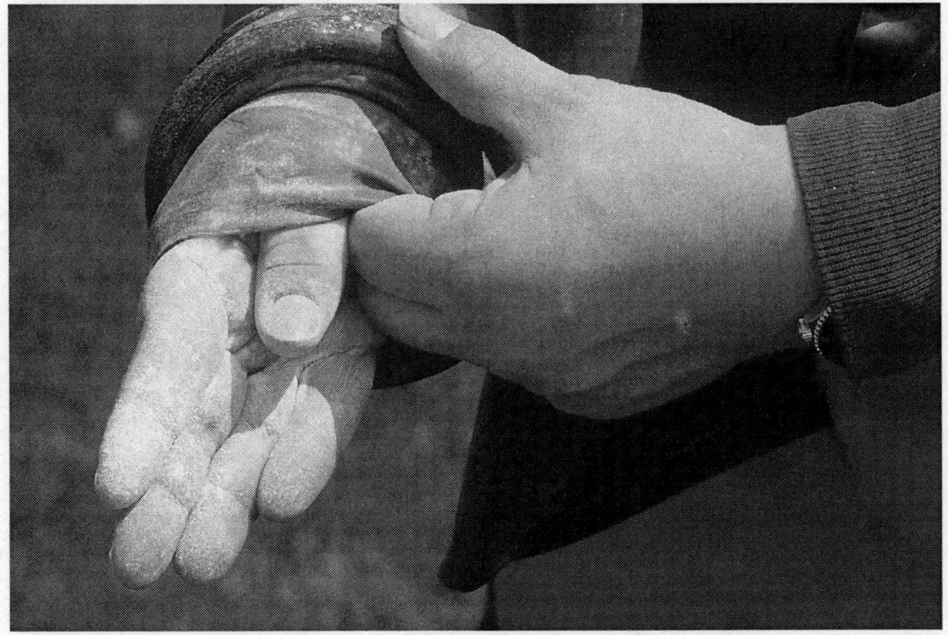

Avoid puncturing the wrist seals with your fingernails.

possible. The tender then places a flat hand against the zipper and carefully uses his other hand to pull the zipper closed. The diver should then give it an extra pull to double-check that it's closed properly. Once the zipper is closed, the diver must burp excess air from the suit. To do this, pull the neck seal slightly away, and kneel first on one knee, then the other. Allow the seal to lie flat again before standing up.

Harness

Once the exposure suit is on, the tender helps the diver get into the harness. To don a harness rapidly, the diver stands with his arms either in a holdup position, level to his head, or with his arms extended straight behind his back. The tender stands

The tender helps the diver get into the harness.

behind the diver and drops the harness over the diver's hands and down his back. The diver then lays one hand flat across his solar plexus as the tender comes around in front of him. The tender puts the chest strap through the first ring, and the diver takes as large a breath as possible as the tender closes the strap through the second ring and gently snugs it down over the diver's hand. This procedure ensures that the tender won't overtighten the harness. It also automatically places the chest strap in the correct position across the solar plexus. For dry-suit divers, this procedure affords some room for air within the suit.

Weight Belt

With the harness in place, the diver is ready to put on the weight belt. However, he should refrain from doing so if it will be some time before he is ready to don the BCD.

The tender should put the weight belt on a diver from behind, holding the free end of the belt in his right hand, and the diver should then buckle it snugly, though not too tightly. If his tender is occupied, the diver can put on the belt himself. To do so, he should likewise hold the buckle in his left hand and the free end in his right hand. This will set up the belt for a right-hand release. The diver then steps over the belt and lifts up both ends so that the belt rests on his lower back. Leaning forward slightly, the diver then tightens and buckles the belt. Tenders should make sure that their divers don't pull the belt too tight, lest they restrict their breathing, possibly without even realizing it. They should also ensure that the buckle has been closed properly. Ankle weights should be snapped in place by the tenders.

If you forget to don your weight belt before putting on your BCD and tank, don't try to cut corners by pulling the belt on under the BCD. It won't be easy to get the belt on properly, and you'll stand a good chance of having it come undone. Take the time to put it on correctly.

Air Supply

All cylinders at the dive site, whether full size or pony bottle, should be laid out according to their readiness. Any full tank should be laid so that its valve faces the water. Tanks that have been used should be placed with their valves facing away from the water. This way, the dive coordinator can assess these resources at a glance. During winter operations, tanks and other gear should never be laid in the snow or on ice, since they may freeze. If you can't leave tanks standing up and secured, lay them on a blanket or tarp.

Anytime a tender touches a BCD/tank/regulator/pony-bottle setup, he should check to see that all of the valves are turned on and all of the other components are functioning properly so that the diver will be ready to enter the water.

Before the diver puts on the BCD and tank, the tender should ensure that the tank is properly secured in the BCD straps. Never pick up the BCD and shake it, because

Full tanks should be laid out so that the valve faces the water.

doing so could cause the tank to drop on your feet. Also, shaking it would only set a loose tank strap at an angle, falsely giving it the appearance of being tight. Instead, hold down the tank valve and the first stage of the regulator with one hand, then try to move the BCD up and down with your other hand. If it moves, the tank needs to be resecured to the BCD. When you check the tightness of the strap, simultaneously look down to see that the tank strap is closed and threaded properly. Make sure that the BCD shoulder strap buckles are closed but extended fully for easy donning. Slip the pressure gauge hose through the BCD armhole so that the gauges can come under the diver's arm and sit securely above the BCD cummerbund in the midchest area.

Immediately after the weight belt is in place, the diver should kneel on one knee with both arms extended behind him and downward. The tender then picks up the tank by its bottom and guides the BCD armholes over the diver's hands. As the tender lifts, the diver stands up, allowing the tender to slide the entire assembly onto his back.

Leaning forward to take some of the weight from the tank, the diver adjusts the shoulder straps, and the tender helps tighten the BCD waist and chest straps. Neither of these should be too snug. Divers typically snug up their BCD chest straps during the gearing-up process with a BCD that is empty or only partially filled. Then, when they're on the surface, they fill the BCD. Since BCDs fill inward, not outward, the result is a tight strap. A full BCD is a squeeze on the diver's chest that restricts his ability to breathe, possibly causing exhaustion or panic. To avoid this, the BCD chest strap should be set so that it's slightly loose with an inflated BCD. Once the strap is set, fasten it with duct tape so that it can't accidentally be pulled too tight on subsequent dives.

The tender takes the weight of the tank off the diver while the diver adjusts the BCD straps.

Gloves

To don gloves, the tender should stand behind and to the side of the diver. The tender puts the diver's elbow against his chest, providing a firm backing to pull the glove on without toppling the diver. The tender can hold the cuff of each glove, allowing the diver to push his hand in. If you have difficulty donning gloves, try blowing into them. The tender should close any zippers or straps. Make sure that the gloves or mitts aren't too tight. Tight gloves will restrict circulation and cause the hands to become cold more quickly.

Tether

Once the diver is ready to walk to the water, the tender should connect the tether to the D ring on the harness with a locking carabiner. The tender can then carry the tether line bag, using the tether as a lead to help the diver up and down slopes, as well as to provide balance on rough terrain. If the diver needs extra balance, he should hold on to the carabiner, not the tether. Holding on to the tether will create slack between them, making balance more difficult.

Mask

Once the diver and tender arrive at the edge of the hot zone, the diver should immerse his face in the water, without a mask on, to acclimate himself to breathing through the regulator. Taking three or four comfortable breaths this way will also serve to prepare him against inhaling water in the event that his mask leaks or comes off during the dive. For divers who use full-face masks, who are more likely to try to breathe through their noses if their mask fails, this procedure is particularly important. If the water at the site may be contaminated, the diver should immerse his face in a bucket of clean water instead. If possible, use cold water to help him acclimate to lower temperatures as well.

Standard masks should be worn under the hood. Before donning the mask, pull the hood back so that it rests around your neck. Put on your mask as you normally would, then pull the hood up and over the mask strap. Wearing the mask this way will greatly reduce the chance of losing it if it becomes dislodged. Full-face masks should be worn over the hood.

Fins

Fins are the last item to go on a diver, and he should never put them on until he's at the water's edge. Walking around in them on land simply poses too much of a risk that he'll trip and fall. Divers may even wish to put their fins on once they're in waist-deep water to make the job easier. If you're diving in moving water, you'll have to don your fins before you go in so that they can be taped to your feet. Otherwise, they could be pulled off by the current.

To don your fins on land, brace yourself against a tender's shoulder with your left hand and with your back to the water. Hold your left fin in your right hand, and cross your left leg over your right leg. The fin straps should be buckled and taped. Once you slip your foot into the pocket and pull the strap in place, reverse everything: Turn around to face the water, brace yourself with your right hand, put your right fin in your left hand, and cross your right leg over your left. Once that fin is on, you'll be ready to enter the water.

Tender-Diver Gear Check

The tender and diver should go through a check of their equipment before the diver is allowed into the hot zone. This check should take less than ninety seconds, and it should be done for every dive.

The tender should have and check the following.

1. An appropriate USCG-approved PFD that is properly closed.
2. Appropriate footwear and exposure protection.
3. Proper tending gloves.
4. An appropriate time keeper.
5. A water-activated flasher.
6. Eye and sun protection.
7. A back-tethered harness if working from a steep embankment.
8. A profile slate and writing utensil if acting as the backup tender.

The tender should check that the immediate dive site has the following five items.

1. A contingency regulator on a full cylinder.
2. A contingency regulator on a full pony bottle.
3. At least one rescue rope throw bag per diver-tender pair if working near fast water.
4. BLS first aid personnel and supplies, at minimum.
5. Potable water.

The tender should go through a head-to-toe check with the diver.

1. The tender should turn on the air valves anytime he touches a tank setup.
2. The hood is in place and not under the mask, unless it is properly sealing a full-face mask.
3. The mask is in place, with nothing under the skirt that might cause leakage. Check that the strap is sitting untwisted across the crown of the head, not over the ears. If the diver is using a full-face mask, check to see that it is secured properly but not too tightly, with no more than two inches of the top strap pulled through the buckle. If a full-face mask is being used without some type of transfer block, check to see that the diver is carrying a backup, normal mask in case he has to switch to a pony bottle.
4. The primary regulator second stage is functional, with the mouthpiece intact and properly secured by a tie wrap. Check the buckle of the tie wrap by wiggling it. There should be no movement. To check the function of the regulator, have the diver exhale through it and then breathe from it a few times. Having the diver exhale first clears any debris from the second stage that he might otherwise inhale. Even more importantly, it helps to program the diver to exhale first, which could be a lifesaving procedure underwater. Simply hitting a purge button isn't enough to test a regulator. A second stage with a missing or damaged diaphragm will deliver air if the purge button is pressed, but it cannot be

cleared and cannot be used normally. Check that all regulator hoses are trim to the diver's body.

5. The pony regulator second stage is functional, with the mouthpiece intact and properly secured by a tie wrap. Again, exhale through the regulator, and then check that it functions properly by inhaling. The second stage of the pony should then be secured in the golden triangle by a mouthpiece cover and snorkel keeper, with the hose under the diver's arm and trim to the body from the first stage.
6. On a dry suit, check that the inflator hose is in place under your arm and secure; that the inflator works; that the suit was properly emptied; and that the vent is properly opened. The neck and wrist seals should have been properly set before the hood and gloves were put on.
7. Ask the diver to press the BCD power inflator three times while the tender monitors the pressure gauge needle, checking to see that it doesn't move. The diver should be able to find the power inflator in one motion and without looking. If the diver has to use his eyes to find the power inflator, he isn't prepared for diving in black water. This procedure ensures that he can find the power inflator, that the power inflator is operational, that the air is on, that the tank

The tender should go through a head-to-toe check with the diver.

is full, that there is air in the BCD, and that the diver knows when to be back on the surface.

8. Ask the diver to reach for each of the following without looking: the primary shears, the secondary shears, the third cutting tool, the carabiner at the harness tether point, and the weight-belt release.
9. Check that the diver is snapped in by a locking carabiner to a marked tether line in a deployment bag. If the diver will be in an overhead environment, ensure that the lock is taped closed with duct tape.
10. Be sure that the webbing of the weight belt extends ten to twelve inches beyond the buckle, that it is clear of the BCD so that it can be ditched, and that it has a right-hand release. Also, check that the webbing is straight in the buckle to guard against accidental opening and release.
11. Ensure that the diver's fins are in hand and properly taped.
12. The ankle weights are in place, if needed.
13. If the diver is to be a backup, he must have a contingency line hooked into the harness tether point carabiner.

Once the tender has completed the second gear check, he and the diver walk to the demarcation line at the hot zone, where they are again checked by the safety officer. All checks should be done thoroughly and expediently so that the diver isn't forced to stand for more than two or three minutes.

Before the primary diver can be deployed, make sure that the backup diver is in place and that the ninety-percent-ready diver is close by. The backup tender should be ready to record the primary diver's cylinder pressure, entry time, and search profile.

The primary tender then assists the diver into the water and ensures sure that the diver acclimates his face to it. The tender checks the pressure one last time. He should also ensure that the diver can reach the first stage of the primary regulator for an over-the-shoulder retrieval.

Study Questions

1. Given the urgency and expectations of an emergency call, it isn't uncommon for divers to become exhausted from arriving and dressing. What are the four basic causes of such exhaustion?

2. Before a diver is asked to suit up, his tender should ask him what two questions?

3. One way of deciding whether the systolic pressure is too high is to add the diver's age to ________ mmHg if the diver is male or to ________ mmHg if the diver is female.

4. Who should close the zipper on a dry suit?

5. At a dive site, how should you lay out a tank that has already been used?

6. On arrival at the edge of the hot zone, how should the diver acclimate himself to breathing through the regulator?

7. During the diver-tender gear check, the tender should ensure that the immediate dive site has what five items?

8. What should a tender do anytime he touches a tank setup?

9. Why isn't simply activating the purge button an adequate test of a regulator?

10. When the diver presses the BCD power inflator, what should the tender see while monitoring the pressure gauge needle?

Chapter 10

The Search

For most entries, divers will simply be able to walk partway into shallow water, put on their fins, and then swim to the descent point. Sometimes, however, it isn't that easy. The best route to the water may be down a steep embankment or off an old pier. Whatever the case, divers should never leap, jump, or make giant-stride entries into low-visibility water. Even familiar waters may contain new, unseen obstructions that could ruin your day on impact. Divers should only enter by walking, descending a ladder, climbing over the side of a boat, or by a controlled, seated entry. Down steep embankments, tenders must belay them by their tethers. From a low dock or off the side of a boat, divers may enter from a seated position, rolling sideways in a controlled fashion and pushing away with both hands. If the diver's feet don't reach the water when he's still in a seated position, then the dock is too high for this maneuver, and he'll have to enter by some other means.

If divers must enter surf, they must be duly trained, and the dive coordinator must agree that the surf is calm enough for safe entries, searches, and exits. Waves enter and break in sets. For the entry, count the wave sets to determine when the large waves roll in, then deploy the divers between those sets. Surf entries should be made by walking backward with all equipment in place, including fins and regulators. The divers should look over their shoulders so as to time the incoming waves and brace against them. Once reaching waist-deep water, or once having been knocked over in shallower water, the divers should turn and start swimming away from shore. When exiting the surf, divers should swim in as far as possible, then crawl on their hands

Divers should only enter the water by walking, climbing down a ladder, or by some other controlled means.

and knees until they are completely clear of the surf zone, keeping all of their equipment in place along the way.

Like giant-stride entries, jumping from helicopters should never be attempted in shallow water, water of unknown depth, or areas that might be concealing obstructions. Beyond the dangers presented by the water, jumping from a helicopter also presents logistical complications. Divers in such operations cannot be tethered, and their backups must either wait on the surface or in the helicopter. Communication with the divers underwater will be limited. If the backups do wait on the surface, the helicopter will have to move away; otherwise, the backups will be beaten by the rotorwash. Finally, not all helicopters are equipped with a hoist to bring a diver back on board. In many cases, a diver in trouble will have to wait on the surface until a boat can reach him.

Tether Length

For many search patterns, a tender has the option of having his diver work his way either toward shore or away from shore. If the search pattern permits, the diver should work his way toward shore. Besides shortening the distance of rescue, it also provides a measure of psychological benefit to the diver, who knows that he is working his way home, even as he becomes colder and more fatigued as the search progresses.

In most cases, a diver should be sent out along the surface to his descent point, but if he is to begin his search fifty feet out on the bottom, then he will have to go out farther than that along the surface, due to the arc involved. The deeper the water, the farther out the diver must travel before he descends. The formula for determining how far out a diver should be sent to compensate for depth involves the Pythagorean theorem, $a^2 + b^2 = c^2$, though the last thing you need to be doing during an operation is geometric calculations. The accompanying table shows the line requirements for reasonable depths. A diver who is to begin his search fifty feet out in water that is thirty feet deep, for example, must be given fifty-eight feet of line. Bear in mind that if the tender is standing more than two feet above the waterline, such as on a dock, boat, or pier, then the height of the platform must be added to the depth.

Surface Distance

Depth	10	20	30	40	50	60	70	80	90	100	110	120	130	140	150
10	14	22	32	41	51	61	71	81	91	100	111	121	131	141	
20	22	28	36	45	54	63	73	82	92	102	112	122	132	142	
30	32	36	42	50	58	67	76	85	95	104	114	124	133	143	
40	41	45	50	57	64	72	81	89	98	108	117	127	136	146	
50	51	54	58	64	71	78	86	94	103	112	121	130	139	149	
60	61	63	67	72	78	85	92	100	108	117	125	134	143		
70	71	73	76	81	86	92	99	106	114	122	130	140	143		
80	81	82	85	89	94	100	106	113	120	123	136	144			

The maximum recommended distance that tethered divers should be sent from shore is 125 feet, and they should never be sent out farther than 150 feet. Line signals don't travel well along tethers longer than that. Even if a communications system is used, line signals are a backup to hardwired radios, not the other way around. Also, the additional drag on the line would lead to diver fatigue and irregular search patterns, and it becomes more difficult for a tender to monitor his diver over greater distances. Perhaps worst of all, distance compromises response time in the event of an emergency. If the team needs to search farther than 150 feet from shore, don't shortcut the operation by giving the diver more line. Instead, switch to a platform-based operation, and continue the search from there.

The Descent

Before the diver launches out, the tender should use a palm-tap signal for the diver to check his air. The diver replies by flashing his fingers, then swims out the appropriate distance. Once the diver has reached the descent point, the tender gives the okay signal, a one-pull tug on the line. The diver quickly verifies that all of his equipment is in place and functioning, then he returns the signal, indicating that he is ready to descend.

Public safety divers should always descend feetfirst. This not only protects their heads from hidden obstructions, it also helps prevent sinus squeeze and allows the ears to equalize more easily. If a diver descends headfirst, the increased blood flow to his head and membranes will make equalization more difficult.

If a diver has difficulty equalizing as he descends, he should stop, remove his hands from his mask, rise a few feet, stop again, and then attempt to clear his ears. The key to this is the phrase "stop again." The Valsalva maneuver (pinching the nostrils and closing the mouth tightly while forcing air into the sinuses and through the eustachian tubes) and other ear-clearing methods involve holding the breath. If the diver attempts to equalize while he's still moving upward, the result could be lung overexpansion injury, especially in shallow water. If a diver experiences any difficulty in clearing his ears, he should abort the dive.

The tender notes the time that the diver left the surface, and he gently controls the line as the diver descends. The profiler should record that information on the recording sheets. On reaching the bottom, the diver gives a one-pull signal, indicating that he has reached the bottom, is comfortable, and is ready to begin searching. The tender returns the signal. All signals are returned.

Searching the Bottom

With the diver in place and ready, the tender issues three pulls on the line *(Go right)* or four pulls *(Go left)*. The orientation of left and right as given here are according to the diver's perspective as he faces back along the tether. The diver's job is to search along the bottom while maintaining a taut line. If there is slack in the line, it's the diver's job to take it up. Thus, if the tender wants the diver to go farther out, the tender simply issues more line.

As the diver searches, he should keep his body at a 45-degree angle to the tether, with his head away from the tender. This position allows him to keep the line taut. It also provides him easy access to the tether to respond to signals, as well as to use his entire body in the search. Keeping the body at a 45-degree angle is one of the most important parts of tethered diving.

As the diver reaches the turnaround point at the end of a sweep, the tender sig-

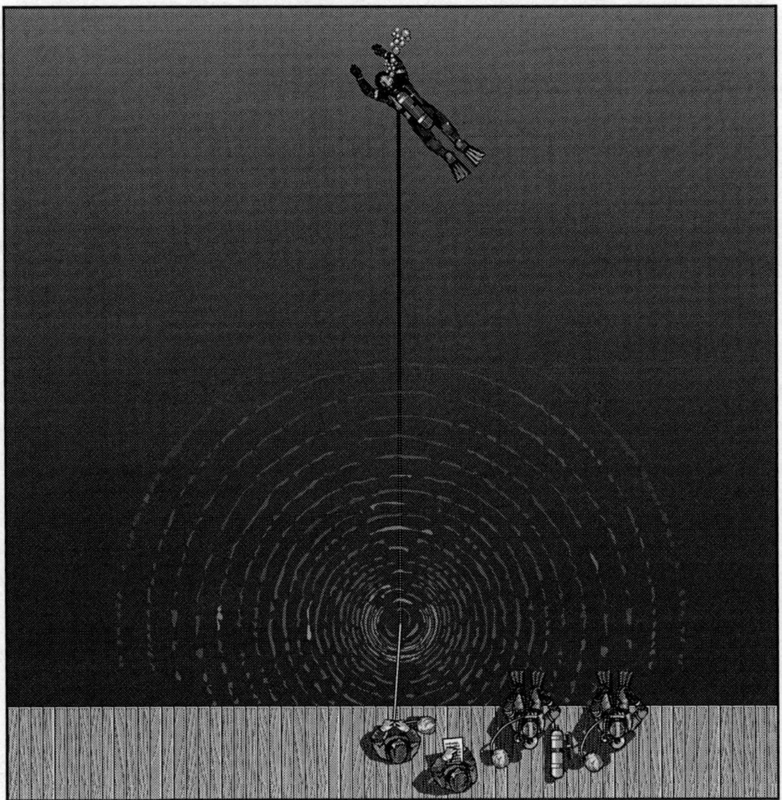

By keeping his body at a 45-degree angle to the tether line, a diver can maintain the tautness of the line.

nals him with one pull on the line, meaning *Stop and face the line.* The diver stops, faces the line, takes up any slack that may have been created, and answers with a single pull. His orientation to the tender and the shore is now refixed in his mind. No matter how dark the water, he'll now have a focal point that makes orientation easier.

As the diver changes direction, the tender brings him into the new sweep by taking in line. The amount of line taken in should be in direct proportion to the visibility of the water and the size of the object of the search.

The demands of a search accentuate the physical demands placed on the body, so a diver might miss what he's looking for because of mental fatigue, physical exhaustion, hypothermia, or simple distraction. His attention span will fall soon after he begins to search. A diver needs to maintain a positive attitude while searching, although the tendency is to do the opposite. Typically, each minute will bring more frustration and disbelief that he'll ever find the object of the mission. Still, he needs to believe that every passing moment is bringing him closer to the target—if it isn't behind him, then it's in front of him. If he doesn't find it, then the next diver will. He must be able to visualize the operation as having a successful outcome.

Mainly because of mental limitations, divers should be limited to a twenty-minute bottom time. When divers are used for such short periods, they're not only more effective, they also tend to be more capable of staying in rotation. In some cases, a diver may be competent and skilled enough that his bottom time can be extended. The decision for a diver to receive a five-minute extension must be approved by the dive coordinator. The principal factors to consider are the diver's experience level; his personal air consumption rate and breaths per minute; the type of exposure suit and the water temperature; whether it is his first dive or a succeeding one; whether the dive is within a safe margin from no-decompression limits; whether the mode of the operation is for rescue or recovery; and whether he'll have to dive again within the next twenty-four hours.

As long as a diver is rested, warm, and hydrated, he may make several searches in the same day. In water of forty feet or less, a diver should have at least forty-five minutes between the first two dives. Additional searches require a surface interval of at least two hours. Depths of greater than forty feet require longer surface intervals.

Any dive operation, especially a rescue, should have as little downtime as possible. Teams should strive for a downtime between rotating divers of no more than two minutes. To do this, the new ninety-percent-ready diver should begin dressing when the primary diver has been down for ten to twelve minutes. He should be ready to become the new backup diver by the time the primary is exiting the water. The current backup, meanwhile, becomes the new primary, deploying to the spot where the original primary left off. The tenders rotate similarly. When taking over, the new primary tender must be sure to stand in the exact spot where his predecessor stood. Since the new primary tender was the preceding backup tender and profiler, he'll know whether the diver's area can be secured or whether it needs to be searched again.

Searching in Black Water

The compulsion to rely on vision as the primary sense for searching is so strong that, even in black water, many divers will swim with their eyes open, peering intently through the mask for a glimpse of *anything*, recognizable or not. The problem that arises is one of fixation. If the diver is able to see anything at all, he'll tend to focus on it. This may happen again and again, and soon he becomes habituated to these shapeless forms, no longer identifying what he sees. This habituation can cause him to miss something even when he looks directly at it.

Two divers in South Africa who surfaced after searching for the body of a child reported that all they found were trees and a big clump of mud. A third diver, however, found the body while he was exiting through the search areas of the first two divers. He later reported that he had used his mind's eye to visualize what he was feeling, and that what at first had seemed like a clump of mud was actually the body of the child.

The sediment stirred up by this diver shows just how quickly visibility can fall to zero. *(Photo courtesy of Pete Nawrocky.)*

A diver searching in black water must use his entire body to search.

For small items, search with only one hand at a time.

Divers searching in black water must use their entire body to conduct a search, visualizing what they feel with their fingertips, arms, torso, legs, and toes. Even a strange brush of the tip of your fin may be an indicator that you've found what you're looking for. In black water, then, you must close your eyes and let your body and your mind's eye do the looking for you.

The technique by which a public safety diver uses his hands is critical. The hands should not grope randomly. Rather, they should have their own search pattern, to be used for every dive. Start with your hands in front of you, palms down. Next sweep them out to each side, then bring them together again. Then, move them forward one hand-length and repeat the sweep. Searching this way ensures that no area along the diver's path will be missed. For small items, such as shell casings, search with only one hand at a time. Hold the other hand in place on the bottom as a landmark for the hand that's searching, and then switch hands to sweep in the other direction. This technique will help a diver cover every square inch of his search area.

Obstacles

Often a diver will encounter a large object in his path, such as an old car, a submerged tree, or rocks. To maintain his search pattern, he must go up and over these obstacles, often at the risk of entanglement. If such an object is interfering with a search pattern, then the area should be divided so that the diver can first search on one side of the object, then the other. If snags remain a problem, you may need to resort to a boat-based operation, allowing you to increase the angle of the tether line.

In monitoring his diver, a tender must ensure that direct-line access is maintained at all times, meaning that there should be no bends, kinks, or corners in the tether. As examples of places where direct access could be lost, a diver should never be allowed to search under a dock on which a tender is standing, nor should he move around to the opposite side of a submerged vehicle. For a search under a dock, the diver may have to be deployed off a boat. For a submerged vehicle, the diver should only cross over the car, never around it.

Discovery

On discovery of the body or missing object, a diver will usually be pumped up or breathing rapidly. Thus, he should wait until his breathing is under control before giving the *Found object* signal. This will decrease the chance of a lung overexpansion injury during the ascent.

Once ready, the diver should indicate that he's found the object by pulling on the tether six times, then six times more. The tender should return the same signal in answer. If the object of the search is a victim, then family members and the media may need to be cleared from the area. The tender should then give a three-pull, three-pull *Stand by* signal to the diver. The tender should then discreetly notify the incident commander, perhaps in code, of the find. At that time, the victim's family can be removed, sheltered, and restrained by law enforcement personnel, if necessary, as EMS personnel get into position to receive the body. Once all is ready, the tender gives a four-pull, four-pull signal to the diver, clearing him to leave the bottom and slowly surface with the victim.

If the diver needs assistance in bringing up the object, he should send a 3 + 3 + 3 signal after the tender acknowledges the 6 + 6 signal, thereby to deploy the backup diver. If a weapon or some other item of evidence is found, it should be marked with a buoy, noted on the profile map, and left for the police investigation and later retrieval. A body should be bagged underwater, especially if it is badly decomposed. As mentioned elsewhere, inflatable lift bags are useful at this juncture, thus relieving the diver of having to handle the body bag all the way to the surface.

Re-Searching an Area

Divers and tenders should be constantly alert to anything that might cause them to miss the object of their search. Divers should signal tenders to make note of objects on the bottom, such as weeds, trees, vehicles, debris, or anything else that may mandate searching the area again.

An area should also be searched again if any of the following conditions apply.

1. A diver shows frequent slack in his tether line.
2. A diver's breathing rate was too high, indicating that he may have been distracted or stressed, and therefore not focused on the search.
3. Divers missed test items that were placed in the search area. These items should be of equal or lesser size than the search object, thus indicating whether the divers are searching thoroughly.
4. The angle of the tether line shows that a diver has come off the bottom during a low- or no-visibility search.
5. A diver has covered too many linear feet. If he is moving too quickly, he may not only be searching too fast, he may also have totally missed an area along the way. This is common. A diver searching effectively in black water for a carabiner-sized object will cover an average of 100 to 300 linear feet during his allotted twenty minutes, whereas a diver searching for an adult body will cover an average of 1,000 to 1,300 linear feet during the same time.
6. A tender is unable to follow correct tending procedures, allowing the diver to take out line, bringing the diver in too quickly, or not using the correct turnaround points.

If the diver isn't holding a good pattern, the tender should be able to see it early and rectify it. Otherwise, the diver should be recalled before the entire span of time is wasted. The success of an underwater operation can be based entirely on one sweep—one diver doing it right, all the way through.

If the search was good but the object still wasn't found, either the object isn't underwater, you're in the wrong location, or the scope of your search wasn't wide enough. While reevaluating all of the information that you have obtained thus far, you can expand the zone of operation. The first approach is to move about twenty feet farther out than your initial sweeps, then begin making sweeps toward the beginning of your first search area. If you still don't find the object, move out another twenty feet, and again search toward the areas that you've previously searched. The second approach is to begin searching outward from the boundaries of the original search. For example, suppose that divers began searching sixty feet from shore on a marked tether line. In expanding the zone of operation, the diver can again be sent out to that distance, only this time, the tender will work the diver away from shore rather than toward it. Because the search zone was logically established in the first place, the object shouldn't be too far from the place you estimated it to be.

Returning

Public safety diving is technical diving. Technical divers use the rule of thirds, meaning that they are halfway through their dive when their tank reaches two-thirds of their starting capacity, and they should be out of the water at one-third. Following this rule, divers starting out with 3,000 psi in their tank should be back on deck with no less than 1,000 psi.

Why use the rule of thirds? Experienced public safety divers will tell you that they often don't realize they have an entanglement until they try to surface. Consider, too, what happens when a diver finds a body toward the end of his dive. What do you do when a face suddenly pops out at you in fifteen inches of visibility? No doubt your breathing rate will go up. If you weren't using the rule of thirds, you'd be at risk of depleting your air supply in a matter of minutes, yet you'd be hesitant to leave the body on the bottom just because you hadn't planned your air consumption well enough.

Because public safety diving is simply more dangerous than sport diving, divers need to use every margin of safety possible.

Divers can surface before being pulled in by their tenders. Doing so allows a diver to control his ascent as needed. If a tender does pull in a diver from depth, he should pull no faster than one foot every two seconds. Any faster may exceed the maximum ascent rate, which might result in decompression sickness or lung overexpansion injury.

A bad habit of many divers is to use the BCD power inflator to rise from the bottom, a method called "riding the elevator." The problem is that, since air within the BCD expands as you move upward, the ascent rate will increase until it becomes too fast to be safe. If you need to inflate your BCD to ascend, it means that you are very overweighted and need to lighten your weight belt.

Still, ascents in black water are difficult. Since you were crawling around on the bottom during the search, you should already have a good idea as to where the surface is. Also, your tether line will show you where the boat or shore is. Take your BCD power inflator in your left hand, hold it straight out, and place a finger or thumb over the dump button, *not* the inflator button. Your buoyancy should be neutral or slightly negative, so start slowly kicking your way toward the surface. If your right hand is free, raise it over your head as a shield against any obstacles. As you rise, use that hand occasionally to feel the angle of the tether line. The change of angle will help give you some idea as to your ascent rate.

As you move upward, the air that was in your BCD to keep you neutral will expand and make you buoyant. Use the finger on the dump button to vent small amounts of air from the BCD. The air that you're dumping wants to go up, and it can only be dumped from the inflator hose if it is above the level of your shoulders.

Continue kicking your way slowly toward the surface. To allow for a gradual decrease of pressure on the body and for the BCD and your lungs to be vented, the currently accepted best rate of ascent is thirty feet per minute, or one foot every two seconds. The importance of taking the requisite amount of time cannot be stressed often enough. If a tender sees that his diver has a habit of suddenly popping into view, he'll know that that diver is ascending too quickly. A tender should actually see the diver's head two or three seconds after his fingertips have cleared the surface.

Once a diver is on the surface and has inflated his BCD, he should indicate to shore personnel that he's all right by tapping the top of his head. On seeing this signal, the tender should gently pull the diver in by the tether line.

The diver should exit the water the same way he entered: in the safest manner possible. The first step for a diver exiting the water is to remove and hand off his weight belt. Tenders should enforce this rule. The fins may be removed next, but all the other gear, especially regulators and masks, should remain in place until the diver is on shore and, if necessary, decontaminated. Leaving the mask and regulator on will protect a diver from contamination, and it will also give him a breathing source and underwater vision if he should somehow fall back in.

As mentioned previously, a tender is responsible wholly to his diver and must stay with him until he has either been dressed down and checked or passed off to EMS personnel. A tender must not abandon his diver to assist with handling the victim or for any other reason.

After a dive, emptied tanks should be filled by qualified personnel. Divers should clean, dry, and stow their gear as soon as possible. Similarly, tenders should clean and dry their own equipment, including tether lines.

Study Questions

1. How should a diver enter surf?

2. Tethered divers should never be sent out farther than ________ feet.

3. How should public safety divers always descend?

4. What is the Valsalva maneuver?

5. Three pulls on the line as issued by the tender tells the diver to go ________.

6. At the turnaround point at the end of a sweep, one pull on the line means to ____________.

7. Divers in low-visibility water should generally be limited to a bottom time of how long?

8. In water of forty feet or less, a diver should have a surface interval of at least how long after the first dive? Additional searches would require a surface interval of how long?

9. Describe the proper technique by which a public safety diver searches with his hands.

10. What should happen if a large object is interfering with a search pattern?

11. Why might the tender use code to notify the incident commander that the body has been found?

12. Name some of the circumstances that might mandate searching an area again.

13. If the target isn't found, either ________, ________, or ________.

14. What is the rule of thirds?

15. What is riding the elevator?

16. Why is it unsafe to ride the elevator?

17. The currently accepted best rate of ascent is ________.

18. True or false: A tender should see the diver's head two or three seconds after his fingertips have cleared the surface.

Chapter 11

Search Patterns

Once the area to be searched has been established, it is up to the dive coordinator to determine what type of pattern will be used. The decision is based on the size of the area, the area from which the tender is working, the available resources, the current, the depth, the object of the search, the probable contour of the bottom, known or expected obstacles, and other such variables.

Divers should be trained to sweep their hands two feet to each side of center when feeling their way across the bottom. Although that is a four-foot span, for the sake of planning purposes, consider from two and a half to three feet to be the average span of a diver's search in black water. That more conservative number helps ensure some degree of overlap between consecutive passes across the bottom. It also provides some insurance against a diver who doesn't quite make full-width sweeps.

In most of the search patterns that follow, the tender controls the latitude of successive passes by way of the tether line, gradually bringing the diver in toward shore. Decreasing the amount that the diver is pulled in at each turnaround point, of course, will increase the overlap of the sweep. In practical terms, the overlap means that even if a diver missed a small object on the first pass, he may still find it on the second. If a tender has difficulty pulling in a diver by three-foot increments, he may opt to use two and a half instead. Since tether lines are marked every five feet, the tender will know he's pulling a diver in at consistent increments by alternately pulling right on a line marker and then exactly between two markers.

In conditions of good visibility, of course, a diver can be pulled in at a greater rate. In water with twenty feet of visibility, a diver searching for an adult body can be pulled in ten feet between each sweep. However, don't make the pull-in distance so great that each sweep is out of sight of the last, or so wide that there is simply too much area for the diver to cover and still maintain a good forward pace. Also, be sure that the search pattern is based on the worst visibility that the diver will encounter. Not only might the visibility drop off on its own, as with increasing vegetation toward shore, the diver might also stir up the bottom on one sweep, ruining the clarity of the water for the next sweep.

When searching visually, divers shouldn't crawl along the bottom as they do in black water. Rather, they should use buoyancy skills to swim over the bottom, scanning from side to side for their target.

Arc Search

The arc search is one of the simplest, most effective, and most frequently used patterns. In an arc search, also known as a half-moon search, the tender remains in

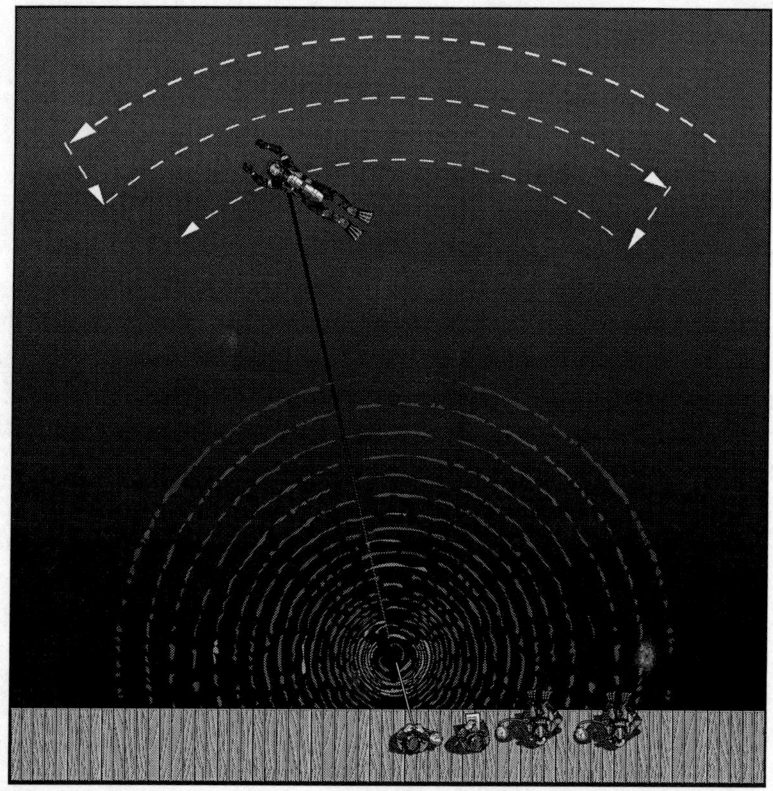

Arc search.

one spot as the diver sweeps back and forth at the end of a taut tether, gradually working his way toward shore. At the end of each pass, the tether pulls in an appropriate amount of line, from two to five feet, depending on the parameters of the search. To be sure that the diver stays within the boundaries of the search area, the tender should pick a landmark on each side of the pattern.

An arc search allows for a rapid search of the bottom, and it is easily used even at the ends of jetties or on a jagged shoreline. It works relatively well over contoured bottoms, and extremely well in searches for both large and small objects. This technique is less than ideal for areas with copious amounts of underwater weeds, however, since the line can become caught in the grass. Not only will this cause the diver to lose his pattern, the weight of the weeds on the line will also be detrimental to safety.

Windshield-Wiper Search

For a faster search of a wide area, arc searches can be combined to create a windshield-wiper pattern. Two primary divers search partially overlapping areas, moving back and forth in unison.

When setting up for this type of search, the tenders should generally stand as far out as the divers deploy. Thus, if the divers are initially sent out sixty feet, the tenders should stand sixty feet apart. A steep bottom or other restrictions might narrow the search arcs and cause the tenders to stand closer together. Keeping the divers moving

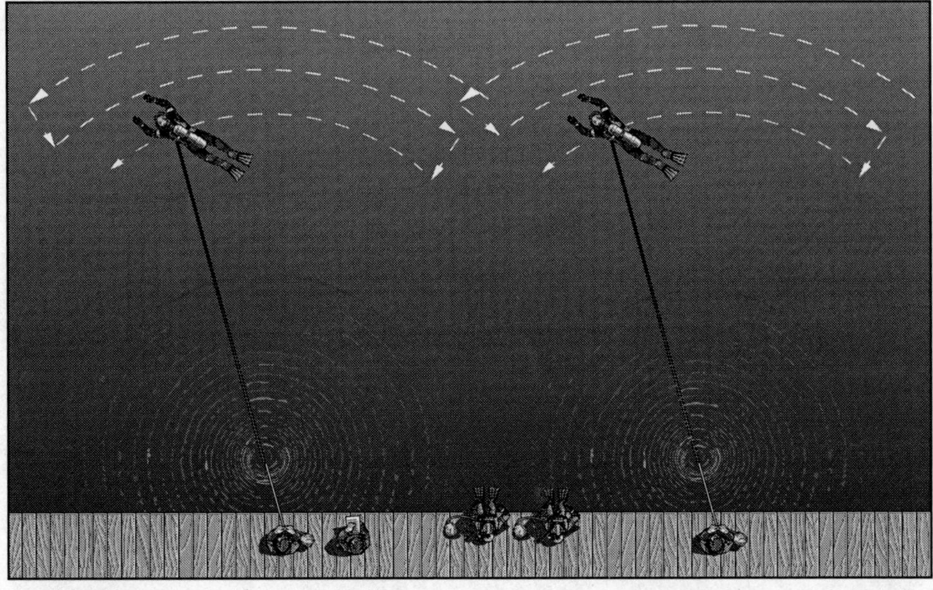

Windshield-wiper search.

at the same pace is up to the tenders. If a tender sees that his diver is getting too far ahead, a single line signal should be sent to tell that diver to stop.

Windshield-wiper patterns allow for an efficient use of personnel. Although both divers have a backup diver and tender, both of those backup divers serve as ninety-percent-ready divers for each other. Thus, for twin search operations, only two other divers need be on shore, but note that both of these must be fully equipped and ready to dive.

If a backup diver is called in by either primary, the other primary should be stopped and, if the situation warrants, slowly brought to shore. Otherwise, there is a risk that two divers may suddenly need assistance and there won't be enough help available. The two primary divers can't act as backups for each other. In an emergency, there simply isn't time for one diver to surface, swim to the other's tender, clip on a contingency line, and descend. Furthermore, it would entail sending someone who has already been on the bottom and no longer has a full tank.

Pier-Walk Search

One requirement of the pier-walk technique is that there must be a long, straight shore, dock, pier, or other area for the tender to walk along. Unlike the arc search, the tender moves along with the diver, creating parallel, straight-line passes

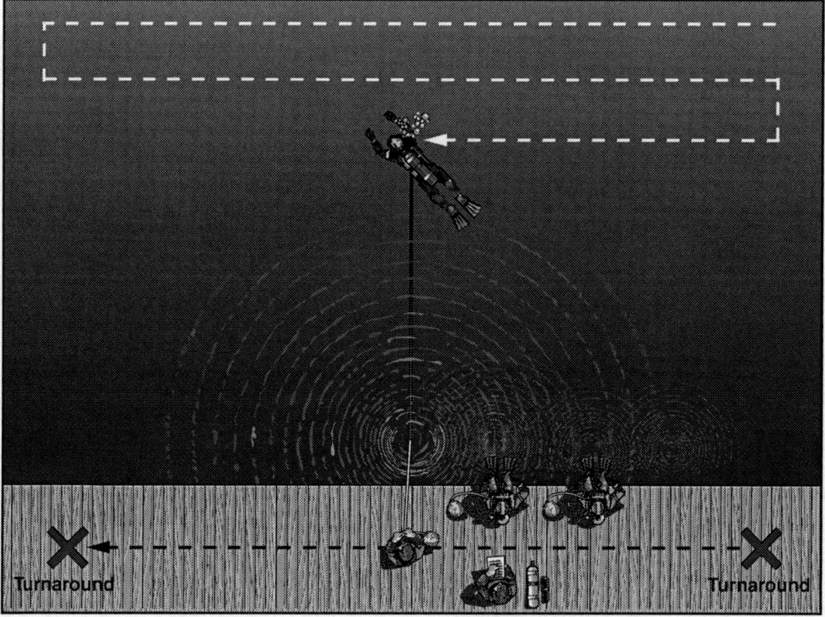

Pier-walk search.

along the bottom. When done correctly, the tender paces the diver as closely as possible, making sure that the tether is perpendicular to the shore at all times. Going too fast or too slow will skew the tether, thereby pulling the diver in too far and causing areas to be missed.

Often referred to as a box search, the pier walk is one of the most efficient, thorough search patterns in use. Because it covers a rectangular area, it doesn't leave pie-shaped sections of the bottom unsearched. It is also a simple pattern, easy to follow and easy to profile. It works well for searching large areas and when operating along featureless shoreline. The pier walk is generally used broadside to a current of up to $1\frac{1}{2}$ knots.

This method of search also introduces an important safety consideration, in that it should be used with a two-person backup system rather than a backup and a ninety-percent-ready diver. Because of the distances involved, the backups should be stationed about one-fourth the distance from each end of the search area. The backups also function as ninety-percent-ready divers for each other.

Double Pier-Walk Search

Pier walks can also be set up as double pier walks, with two divers searching at distances. Typically, one diver works twice as far out as the near-shore diver to pre-

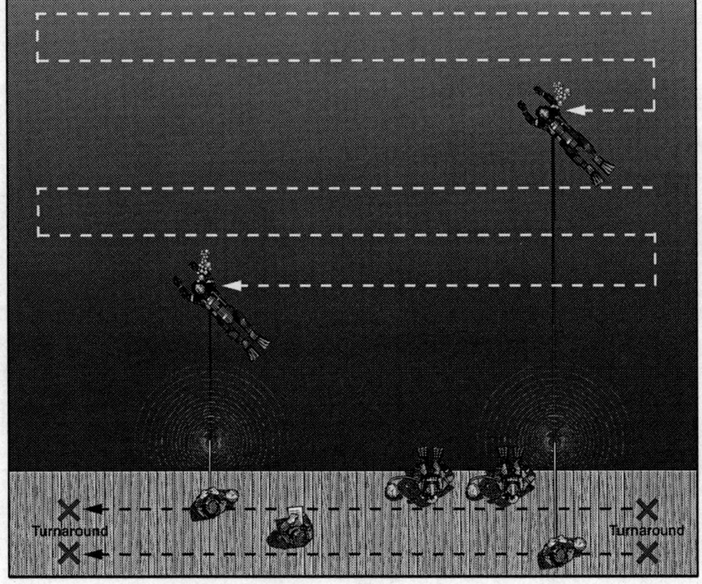

Double pier-walk search.

vent their lines from tangling. The tenders walk separate, parallel paths, and one ducks under the other's line as they pass. The divers need not work in unison, but two backup divers should stand by as in a windshield-wiper pattern.

Jack-Stand Search

The jack-stand technique lends itself well to bottoms that are nearly flat and for searching for small items, such as weapons, evidence, and personal property. It is meant for a methodical search, rather than one covering a large amount of area in a short time. This technique can be used with a single diver who is tethered back to the surface, or it can be used with two divers and no tether. It can also be used perpendicular to a light current (one-fourth to one-half knot) or parallel to a current of up to one knot.

The jack stand involves using one running line on the bottom. The running line, weighted with fifteen to twenty-five pounds at each end, should be no more than 125 feet in length. A vertical line to the surface, carrying a float of at least ten pounds buoyancy, is attached to each weight.

The diver descends to one end of the weighted running line. On reaching the other end, after searching methodically along its length, he moves the weight two and a half to three feet, or as visibility and the size of the object in question dictate. Whenever there is a current, even a slight one, the line should be moved upstream. This way, the silt that he raises on each pass won't settle over the search area.

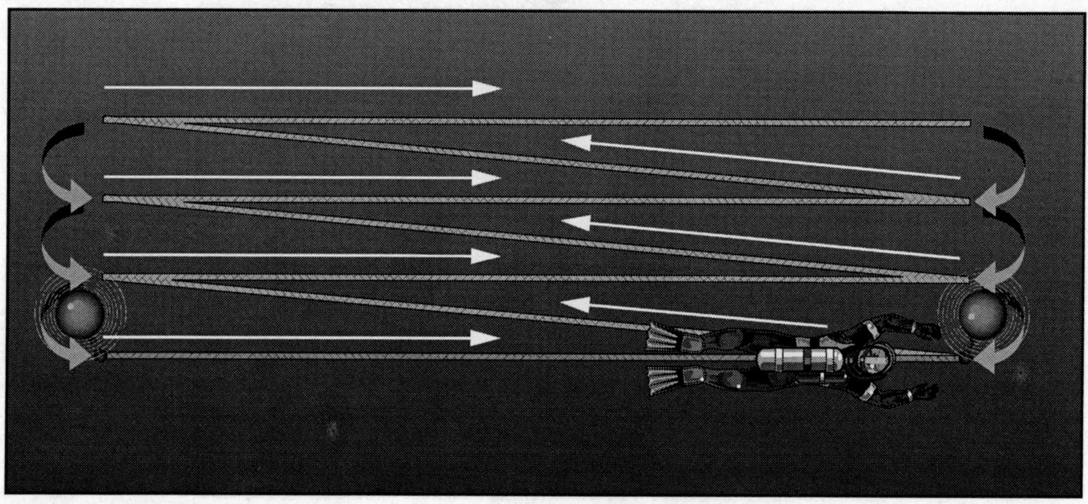

Jack-stand search.

The main problem with a jack-stand pattern is that it works poorly with a tethered-diver system, since the diver is too likely to become tangled in all the attendant lines. Without a tether, the significant risk in coming off the line is that the backup won't be able to find the primary in a timely manner. There is also a greater risk that the diver will surface under an overhead environment or in boat traffic. A quick-release contingency line can be used to connect the diver to the running line to help forestall these eventualities. Lacking a direct connection to shore, a jack stand requires that two divers be deployed rather than one, decreasing the efficiency and effectiveness of the search. Jack-stand divers typically keep one hand on the line, further limiting their capability. Using this method, it is also very difficult for the tenders to know and document exactly which areas have been searched and which have been missed, even with buoys deployed from each weight.

One more problem that divers often encounter when using this pattern is that the running line along the bottom can become snagged on debris when the diver moves a weight forward. If a diver finds such a snag as he searches along the line, he'll have to unsnag the line, pull it taut, and search that particular sweep again. The weights that anchor the ends of the line must be heavy enough that a diver can pull against them to tauten the line.

A common reason why teams and trainers choose the jack stand is because they weren't trained in how to anchor a boat effectively for tethered diving. If such is the case, then a jack-stand search is certainly preferable to a free-dive method. Wireless underwater communications systems are extremely useful for this type of search. If your team doesn't use such a system, run a signal line from one of the buoys to shore.

Circle Search

As the name implies, a diver following this technique makes a circular pattern over the bottom while tethered to a central point. A second diver can sit at the hub to let out line as needed, or the tether can be rigged through a weighted swivel, with the line run up to a tender aboard a boat. A third method is to use a MacKin pivot, which is a weighted unit with a free-moving spool, allowing a diver to make unimpeded circles around it. In this case, the diver lets out his own tether line from a spool attached to his harness. Whenever a MacKin pivot is used, it should have a buoy line to the surface, indicating the center of the search pattern and also to act as a signal line.

If a diver is controlling his own search, he must place a marker on the bottom so that each time around he'll know when extend the search. Also, if there is no tether line back to the surface, then a second diver must be in the water with him to act as a buddy.

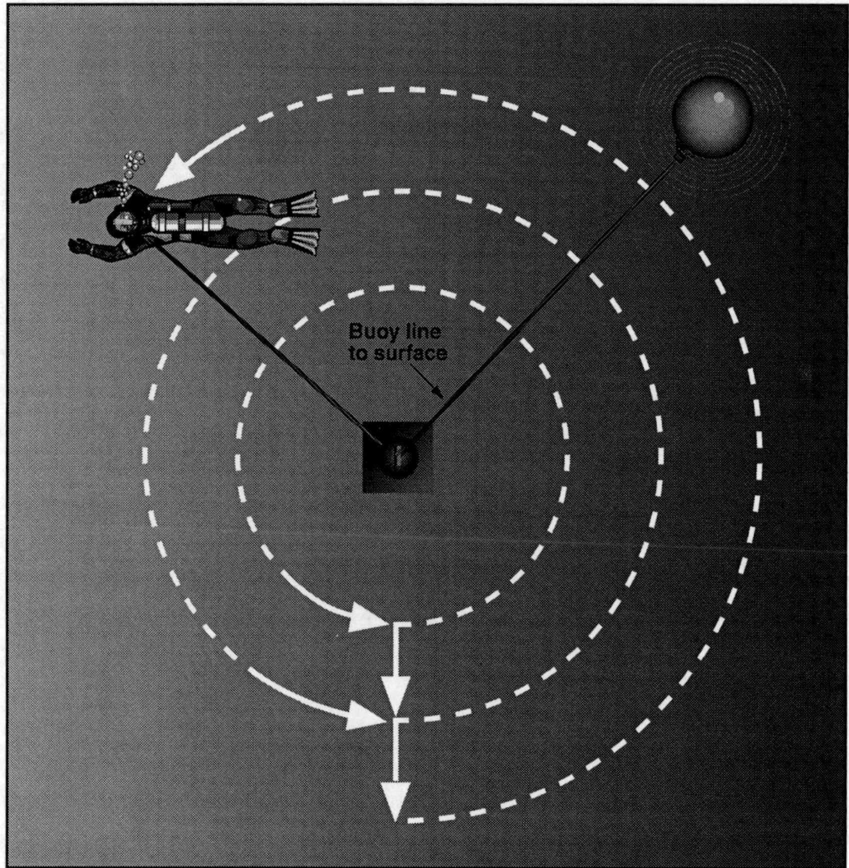

Circle search.

Rather than swimming with the primary, the buddy should remain at the hub, where he can monitor and send line signals.

Circle searches aren't usually very effective for several reasons. First, they pose a risk of entanglement, no matter which arrangement of divers and lines is used. Second, these patterns aren't directed by the tender. Conducting multiple searches this way will leave large unsearched areas outside the perimeters unless the circles are well overlapped. Finally, the worst problem is that there is little way to map the area actually searched.

Running-Line Search

The running-line search is extremely useful in areas of heavy grass, debris, or strong currents. The diver searches straight out from shore or from a boat to the extent of the determined search pattern. If the water is moving, he deploys with the current. Once he reaches the end of the first sweep, he surfaces, then he and the tender move as a unit anywhere from two to five feet to one side, depending on the parameters of the search. The diver then searches toward the tender, and the process repeats itself. This technique affords a reasonable degree of control while reducing somewhat the problem of snagging weeds. If the problem is debris, the diver will be able to crawl over it and untangle himself on his return, thereby minimizing the areas that he misses. Deploying with and returning against the current means that the diver won't be pushed off of his search pattern. Backup divers are used as normal.

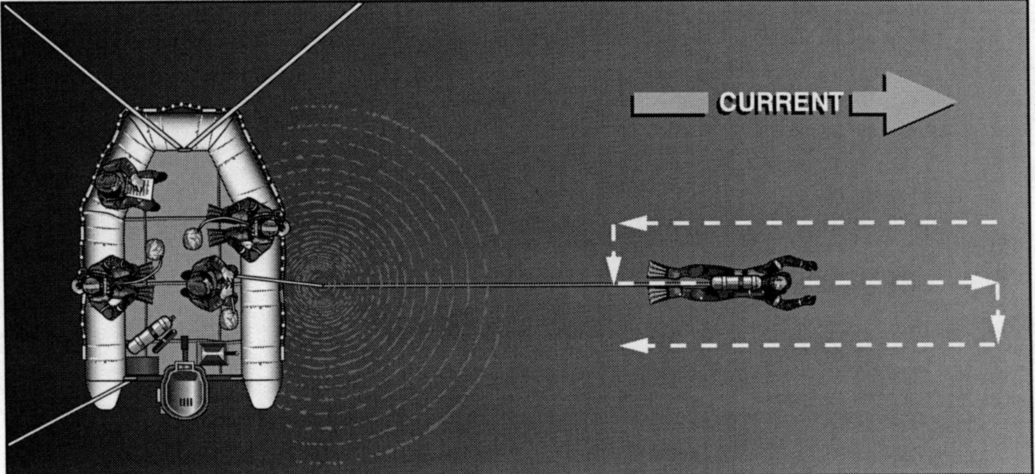

Running-line search.

Direct Overhead Search

In areas with exceptionally heavy weeds or debris, you may need to use a direct overhead search. For this pattern, the diver descends straight below the tender, who is on a platform, and searches the immediate area. The diver is then raised above the weeds and lowered again a foot or two to one side. If you must use this type of search, be sure that the divers aren't pulled up too quickly by their tenders. The repetitive up-and-down motion greatly increases the risk of ear and sinus barotrauma, as well as lung overexpansion injuries.

Direct overhead search.

The direct overhead search is used in areas of heavy weeds or debris, as around these old pilings. *(Photo courtesy of Frank Pagliardi.)*

Frame Search

Searching for an extremely small item, such as a ring or bullet casing, can be tedious and difficult. Even with a tender, a diver may have a hard time ensuring that he has searched every inch of the bottom. The solution is to build a square frame of wood, metal, or plastic to use as a small arena for a concentrated search. Weight the frame, if necessary, so it'll stay on the bottom. Lay it down at the beginning of the first sweep of a normal search pattern, then meticulously search within its confines and up to a hand's width around its outer edge. If the search needs to be truly thorough, use a sieve in combination with the frame. When you're done with that area, lift up the side closest to you and, leaving the opposite side on the bottom, flip the frame over so that it covers a new area adjacent to the one just searched. At the end of a sweep, flip the frame laterally. Obviously, this search method works best in combination with a pier-walk pattern, which, like the frame, covers a rectangular area.

When a primary's bottom time has been exhausted and the search needs to be passed to the next diver, he should simply leave the frame on the bottom. Because the team is using tether lines and profiles, it's easy to send a replacement down to resume the trail. Divers should not forget to search under the frame itself.

Free Search

Free searching should almost never be used in waters of low visibility, not only because it is inaccurate and inefficient, but also because of the danger it creates for divers. If a diver has some type of problem, such as entanglement or air depletion, he has no way to signal for help, and there is no way to find him immediately, especially in black water. The only times that public safety divers should use free search are when an extremely restricted area, choked by trees or other debris, would continuously hinder the movement of a tethered diver, and in very fast water. In the first instance, free search requires reliance on the buddy system, and both divers must carry a fully redundant pony bottle. This is most commonly done in limited areas, where the diver cannot wander off too far. In the latter case, in currents of more than three knots, the diver deploys, drifts across the bottom, and is retrieved after surfacing by trained personnel. The diver is then taken upstream to begin again. Fast-water drift searching requires a highly trained surface support crew and meticulous moment-to-moment record keeping if the risks of diver loss or injury are to be minimized.

Combining Patterns

In choosing which pattern is the most appropriate for a given location, you aren't limited to a single type of search. Different patterns can be combined to make the operation faster and more efficient. In an operation off a pier, for example, it may be appropriate for the tender to guide the diver along a pier-walk pattern, then allow him to swing out and around in an arc pattern at the far end.

Multiple-Diver Line Searches

Some teams opt for a search system common to sport diving, in which several divers, usually four to six, hold on to a single line that is anchored either to shore or by a person underwater. The divers then swim along as a unit, covering large amounts of territory in a single pass. Although the plan sounds good, it has so many pitfalls that it often quickly deteriorates into a scene of confusion. Perhaps its foremost problem is that there are simply too many people to coordinate. Divers often experience difficulty in weighting, maintaining neutral buoyancy, equalizing, breathing, and all else. If one searcher along the line experiences such a problem, the entire sweep is delayed. Also, when divers are trying to hold on to a line, keep pace with the other divers, and pay attention to a multitude of signals, much of the physical and mental focus that should be devoted to the search itself is instead directed toward swimming as part of a coordinated machine. Obviously, not all divers swim at the same rate, and this in itself poses difficulties. If one diver encounters a debris pile, there is no way for him to tell the others to slow down so that he can search it properly. Even if he could, that would slow down the other divers unnecessarily. Even when everything is nominal, holding on with one hand means that, at best, each diver is searching only half of what a tethered diver can search in the same time period.

Finally, this method is problematic for emergency situations. Connected to a single line by their harnesses, one diver may be prevented from making an emergency ascent. On the other hand, if the divers merely hold on to the line, they can become disconnected just by letting go. A diver in distress would have only one hand free for self-rescue. If one were to become entangled along the way and drop the line, how would the others know it? Worse, the divers could become entangled all at once. If the group were to swim into a fishing net, for example, they wouldn't be able to act as backups for one another, and their signals to shore could become confused with so many divers pulling on the line. Properly, using this broad-sweep technique would demand having a backup diver for every member who deploys underwater. Given the almost universal limitations of staffing, this would be impossible for most teams.

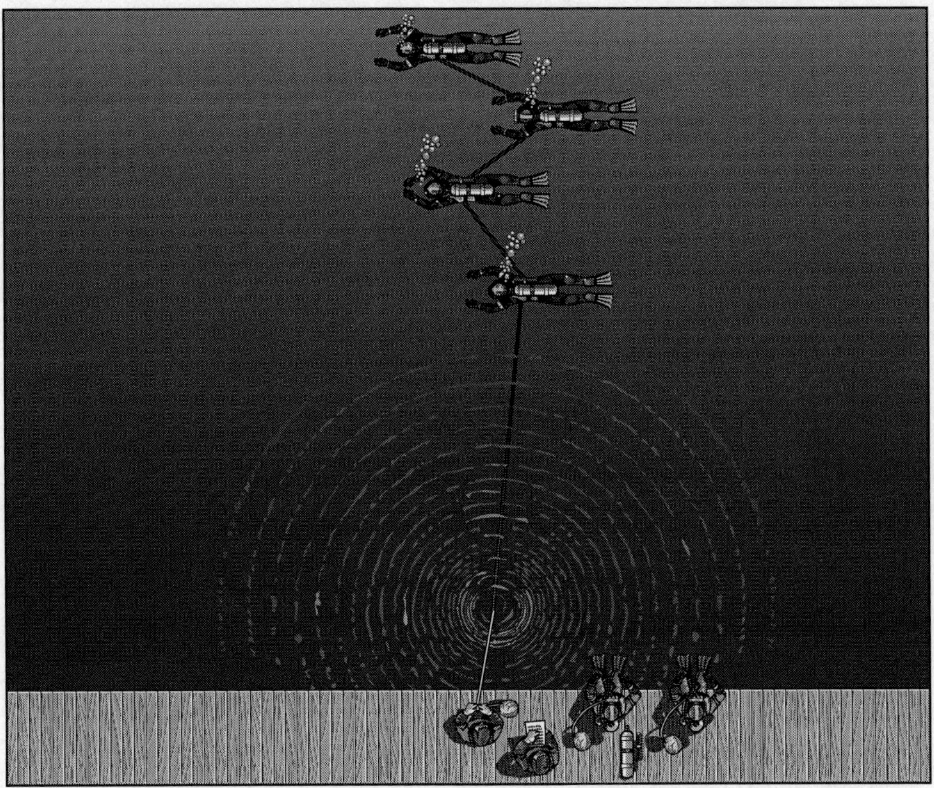

Deploying search divers on a single tether poses a multitude of problems, not the least of which is coordination.

Walking Search

Not all underwater searches have to involve scuba gear. If the water is shallow; i.e., waist-deep or lower, your team may opt to conduct a walking search rather than dive operations. In a walking search, one or more personnel walk in search patterns, literally in an attempt to stumble over the target. Although this method is unsuitable when searching for small items, it can be quite effective when looking for something larger, such as a human body.

In a walking search, there are two ways to maintain the pattern and ensure that the area is canvassed properly. First, the walkers can be tended in the same manner as divers. Second, the walkers can simply use landmarks and visual aids as a guide. Multiple walkers can be effective. By holding hands and staying close together, a team of walkers can cut a wide swath through the search area.

A few safety considerations apply to this type of search. Don't use this technique in water higher than the waist, and don't deploy walkers in a current of more than two knots. Faster water may knock over personnel and drown them. Also avoid performing a walking search in rough surf. In any type of water, all walkers must wear PFDs over their exposure suits because of the possibility of tripping or getting a foot caught. Rocks, debris, and muddy bottoms can make walking searches difficult or impossible. If personnel trip, fall, or get stuck in the mud, continue the search using divers. Keep numerous rescue rope throw bags ready to assist any searchers in the water. Also keep a backup diver at the ready, for even in the shallows, a diver may be required to liberate a walker from a severe entanglement.

Shotgunning

Imagine yourself at a rescue site for about fifteen minutes without any sign of the victim. You start wondering whether you're in the right place, and over the next couple of minutes, your doubts grow stronger and stronger. A little voice starts yammering inside your head, telling you that you should change sites.

If you really are looking in the wrong place, you wouldn't be the first to do so, but what if you were right in your original assessment and you choose to look elsewhere? Shotgunning, or deploying divers to swim around underwater with no plan and no pattern, occurs when you don't have enough faith in your plan to stick with it.

If you find yourself wanting to change sites because of a growing doubt, consider whether your hunch is really a better indicator than witness information and good profiling techniques. If your team is doing a proper search in the area where you have objectively determined the victim to be, then you shouldn't let your doubts get the better of you. Finish searching that area before moving on. Otherwise, you'll simply jump from site to site, never completing a search and setting yourself up for failure.

Calculating the Area Searched

A diver surfaces after searching for a small object in black water. The profiler, who will become the next primary tender, needs to decide whether that area has been searched properly or whether it needs to be covered again. Documenting how large an area has been searched is an important step in making this decision. Knowing how much area an average diver can accurately cover is also important for calculating the number of divers needed for a particular area within a set amount of time.

For rectangular patterns, calculating the area searched is fairly straightforward.

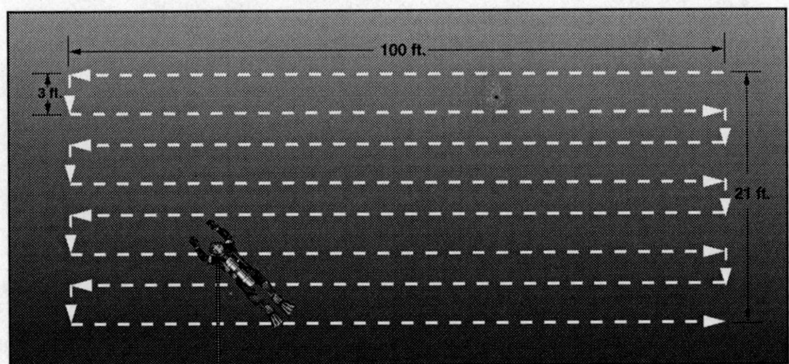

The search area of a rectangular pattern can be calculated by multiplying the lateral distance by the range of deployment.

Suppose a given pier-walk pattern, for example, involved a lateral distance of 100 feet. Suppose, too, that the diver was originally deployed to a distance of 60 feet and that, when the tender brought him out at the end of his twenty minutes underwater, the tether line was showing 39 feet. 60′– 39′ = 21′. Multiplying that figure by the lateral distance, 100′, gives you the area of the rectangle searched, or 2100′. (These calculations will discount the additional span of the diver's outer arm along the fringes of the search pattern, worth about a foot and a half. For the mathematically fussy, this quantity may also be factored in, or it may simply be considered negligible overlap with the adjacent pattern.) Counting the number of passes that the diver made and multiplying that by the lateral distance will give you an idea as to the diver's efficiency during the operation. For example, if the diver made eight complete sweeps during the operation described above, then 8 x 100′ = 800 linear feet, representing the actual distance that the diver covered. On a twenty-minute dive, an average diver can accurately cover 100 to 300 linear feet when looking for a handgun, 500 to 600 linear feet when looking for a rifle or shotgun, and 600 to 800 linear feet when looking for an adult body. The figures given are for black water and in the absence of either heavy debris on the bottom or a notable current. If the bottom is flat and free of debris, the maximum that a black-water diver can cover is 1,200 linear feet in twenty to twenty-five minutes. Faster rates typically result in outrunning the line, which results in slack and a poor pattern.

Since arc searches involve curves, calculating the area covered is a bit more complicated, although not intimidating. Suppose a diver was originally sent out to a distance of 100 feet and was brought out of the water at 70 feet. Suppose, too, that his search covered an arc of 90 degrees, or one-fourth of a circle. The area searched will be equivalent to the area of a circle with a 100-foot radius minus the area of the inner, 70-foot circle, divided by 4. Since the area of a circle equals $\pi \times \text{radius}^2$, and using 3.14 for π, the math works out as follows.

3.14×100^2 = area of entire 100-foot circle, or
$3.14 \times 10{,}000 = 31{,}400$ sq. ft.

minus...

3.14×70^2 = area of 70-foot inner circle not searched, or
$3.14 \times 4900 = 15{,}386$ sq. ft.

or...

$31{,}400 - 15{,}386 = 16{,}014$ sq. ft.

divided by 4...

$16{,}014 \div 4 = 4{,}003.5$ sq. ft. = area searched.

In applying this formula in the real world, the greatest room for error, unless measured precisely, will likely be in estimating the arc of the sweep. Again, these calculations do not account for the span of the diver's outer arm along the fringes of the pattern.

Determining the linear feet covered by a diver in an arc pattern is a matter of adding the circumference data for each successive tier of the search, again, calculated as a portion of an entire circle. If the diver in the above example made seven complete passes and was brought in five feet at each turnaround point, adding up the dis-

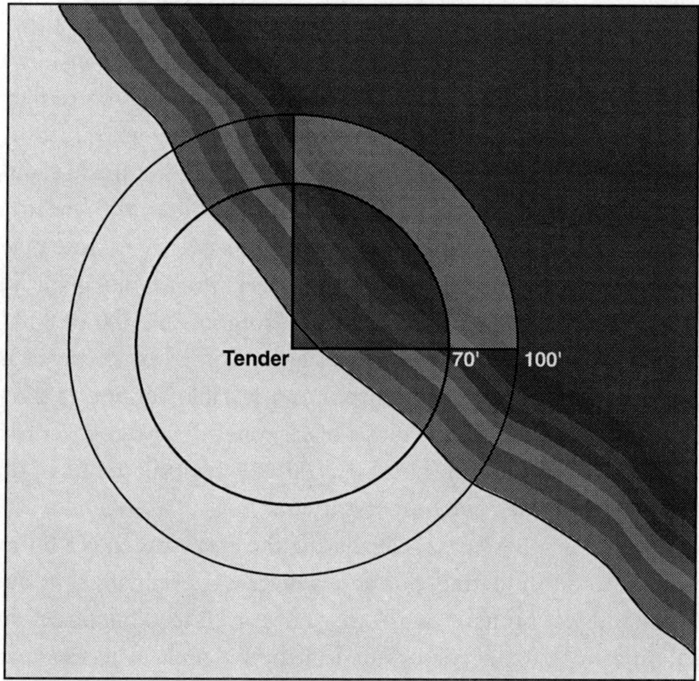

The area covered by an arc pattern can be determined by calculating the overall theoretical search zone, based on the length of the tether, minus those sections that were not searched.

tance covered by each of those passes will render the overall linear feet.

Since the circumference of a circle is equal to $2\pi r$, the distance covered along the seven sweeps of his search is determined as follows.

At 100 feet: 2 x 3.14 x 100 = 6.28 x 100 = 628 ft. ÷ 4 = 157 ft.
At 95 feet: 2 x 3.14 x 95 = 6.28 x 95 = 596.6 ft. ÷ 4 = 149.15 ft.
At 90 feet: 2 x 3.14 x 90 = 6.28 x 90 = 565.2 ft. ÷ 4 = 141.3 ft.
At 85 feet: 2 x 3.14 x 85 = 6.28 x 85 = 533.8 ft. ÷ 4 = 133.45 ft.
At 80 feet: 2 x 3.14 x 80 = 6.28 x 80 = 502.4 ft. ÷ 4 = 125.6 ft.
At 75 feet: 2 x 3.14 x 75 = 6.28 x 75 = 471 ft. ÷ 4 = 117.75 ft.
At 70 feet: 2 x 3.14 x 70 = 6.28 x 70 = 439.6 ft. ÷ 4 = 109.9 ft.
Adding these distances gives a total of 934.15 linear feet.

If math isn't the forte of the public responder, the following table, rounded to the nearest foot, will allow you to make rough calculations easily. The linear-foot distances are all based on an arc of 90 degrees.

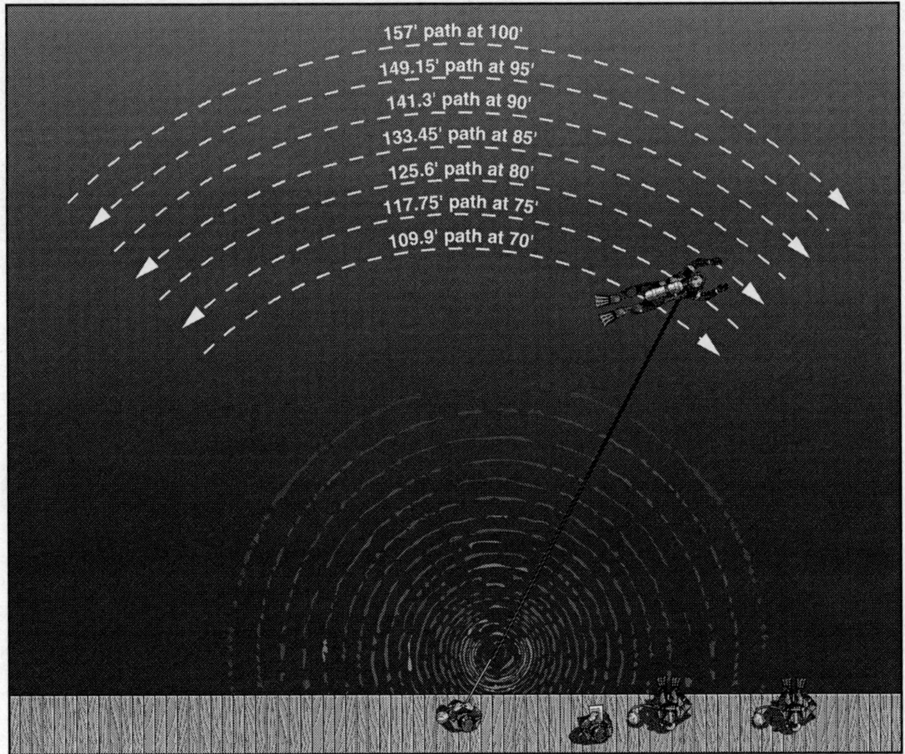

Determining the linear feet covered by a diver in a arc pattern is a matter of adding the distances covered in each sweep.

Distance from tender	Linear ft. (90° arc)	Distance from tender	Linear ft. (90° arc)	Distance from tender	Linear ft. (90° arc)
150	236	100	157	50	79
149	234	99	156	49	77
148	232	98	154	48	75
147	231	97	152	47	74
146	229	96	151	46	72
145	228	95	149	45	71
144	226	94	148	44	69
143	225	93	146	43	68
142	223	92	145	42	66
141	221	91	143	41	64
140	220	90	141	40	63
139	218	89	140	39	61
138	217	88	138	38	60
137	215	87	137	37	58
136	214	86	135	36	57
135	212	85	134	35	55
134	210	84	132	34	53
133	209	83	130	33	52
132	207	82	129	32	50
131	206	81	127	31	49
130	204	80	126	30	47
129	203	79	124	29	46
128	201	78	123	28	44
127	199	77	121	27	42
126	198	76	119	26	41
125	196	75	118	25	39
124	195	74	116	24	38
123	193	73	115	23	36
122	192	72	113	22	35
121	190	71	112	21	33
120	188	70	110	20	31
119	186	69	108	19	30
118	185	68	107	18	28
117	183	67	105	17	27
116	182	66	104	16	25
115	181	65	102	15	24
114	179	64	101	14	22
113	177	63	99	13	20
112	176	62	97	12	19
111	174	61	96	11	17
110	173	60	94	10	16
109	171	59	93	9	14
108	170	58	91	8	13
107	168	57	90	7	11
106	167	56	88	6	9
105	165	55	86	5	8
104	163	54	85	4	6
103	162	53	83	3	5
102	160	52	82	2	3
101	159	51	80	1	2

Again, the two main uses of this information are in deciding whether a diver has moved at a reasonable rate when executing his pattern, and in figuring out how many dives are needed to cover a given search area. If a diver searched 500 linear feet for a handgun in zero visibility for twenty-two minutes, for example, you know that the area needs to be re-searched, even if the diver held a taut pattern. To have covered that much distance, the diver was definitely moving too fast and could have easily missed the target.

Consider also a team that needs to search low-visibility area of 500 feet by 250 feet (125,000 ft.^{2}) for the body of an adult. The commander wants to know how long this will take and how many divers will be needed. Using 800 linear feet as an average that each diver can cover in each dive, multiplied by an average 3.5-foot sweep, gives a result of 2,800 square feet per diver. The 125,000 ft.^{2} area divided by 2,800 sq.^{2} equals 44.6 dives (call it 45). With twenty-five-minute dive times, the search will require 18.75 hours with one diver searching, or about 9.4 hours with two divers down simultaneously. If five divers perform four dives, and five other divers perform five dives, then this mission can be ably handled by a total of ten divers.

STUDY QUESTIONS

1. Name some of the factors that the dive coordinator will consider when deciding what search pattern to use in a given area.

2. Divers should be trained to sweep their hands how far to each side of center when feeling their way across the bottom?

3. How should a tender keep the diver within the turnaround boundaries of the search area when using an arc search?

4. When using a windshield-wiper search, how far apart should the tenders stand?

5. How many backups are required for a pier-walk search?

6. How many ninety-percent-ready divers are required for a pier-walk search?

7. Must the divers in a double pier-walk search work in unison?

8. Under what circumstances might a jack-stand search be of benefit?

9. A weighted device with a free-moving spool that allows a diver to make unimpeded circles around it, as in a circle search, is called a ________.

10. True or false: In a running-line search, the diver deploys across the current.

11. In areas with exceptionally heavy weeds or debris, it is probably best to use what type of search?

12. When performing a meticulous search for an extremely small object, it is probably best to use what type of search?

13. During a twenty-minute dive, an average diver can cover ________ to ________ linear feet when looking for an adult body in black water, unimpeded by debris.

Chapter 12

Communication and Line Signals

The best option for communicating underwater is to use an electronic system. When functioning properly, such a system reduces the chance of miscommunication and ensures that the tender always knows the status of the diver. Being able to talk to someone is also psychologically reassuring for a diver, especially during a body recovery.

Communications systems, available commercially, can be either hardwired or wireless. A hardwired system relies on a cable that physically connects the diver's gear to a unit on the surface. Besides the tender's microphone and speaker, some units also accommodate connections to other divers' units and recorders. Hardwired systems generally provide clear, strong sound, and they're usually less expensive than wireless systems. Since the divers are tethered anyway, teams might as well work with the system that provides the clearest communication.

A wireless system uses sound waves to transmit signals through the water. Although some types transmit and receive signals well, all wireless systems are prone to background noise, such as that caused by boat engines, bubbles emitted by the diver, and air passing through the diver's regulator. Acoustical considerations also come into play, since ships hulls, rocks, and concrete piers can all reflect and distort signals. Although wireless systems do offer the advantage of allowing a diver to remain in contact even if he has to cut loose from his tether, abandoning the tether is a dangerous, unlikely scenario—one that shouldn't predicate whether to purchase a given system or not. All things considered, a wireless system is the best choice for teams that operate in high-visibility water, performing tow searches and untethered patterns.

Both wireless and hardwired systems require either a full-face mask or an oral cup to allow the diver to speak. Before using either type of setup, be sure that you are trained in its use and emergency procedures. Voice-activated systems are preferable to push-to-talk varieties, since the diver won't have to stop searching with one hand whenever he needs to communicate.

Underwater communications systems are especially important where direct-line access between the diver and tender could easily be lost, such as when diving under ice or other overhead environments. Even so, the basic system of communicating should still apply. It's easy just to tell a diver to "Go left" or "Go right," but it's far wiser to say, "Go three" or "Go four," since doing so will reinforce the use of basic line signals. If the technology fails at any point, the diver and the tender can still resort to pulling on the tether, but their proficiency at this will hinge on their experience and recency of practice.

Line Signals

Anytime a pull signal is given, it must be acknowledged with a return of the same signal, thereby to ensure that no miscommunication is taking place. When giving multiple signals, wait for acknowledgment of the first before sending the next.

Tender-to-Diver Signals

One pull: Stop, face the line, take up the slack, and prepare for a new signal. By stopping to face the line, a diver can reorient himself toward shore. This mental picture fosters an opportunity to move backward and take up any slack in the tether. The diver signals that he is okay and ready by returning the single pull. In general, a single pull at any time from the tender means, "Are you okay?" and a single pull returned from the diver means, "I am okay." All line signals from the tender should be preceded by this one-pull signal.

Three pulls: Go to the tender's left, diver's right. If by some chance this signal is misunderstood and the diver heads in the wrong direction, the tender sends a one-pull signal. The diver stops, faces the line, and receives another three pulls. Reoriented, the diver returns the signal and proceeds in the correct direction.

Four pulls: Go to the tender's right, diver's left. For a memory trick, face the water, then hold up your left hand with your fingers spread and the back of your hand facing you. The four fingers stand for the four letters in the word "left," and your thumb is pointing in the direction that the diver ought to go. If you, as a tender, ever have trouble remembering the signals, put three marks on the back of your left-hand glove, and four marks on your right-hand glove. Even adults are known to confuse left and right, but a search area is no place to be mixing up directions.

Two plus two pulls: Search the immediate area. As always, the diver should return the same signal as acknowledgment. With this signal, the tender is requesting the diver to perform a slower search of the area directly in front of him. It does not mean that the diver should move out of the pattern.

Three plus three pulls: Stand by and prepare to leave the bottom. This signal is used to tell the diver that there may be a change in the operation. Perhaps new witness information has been received that will change the search area. The diver should stop on the bottom and wait for another command. This signal is used when the target has been located and shore personnel need the diver to stand by while they clear out the family and media, or make other preparations. When conducting multiple searches simultaneously, if one diver issues a help signal, all of the other divers should immediately be put on standby with a 3 + 3 signal.

Four plus four pulls: Surface. For this signal, tenders and divers may wish to agree ahead of time that the diver won't return this signal when he receives it; otherwise, he'll only acknowledge it with a single pull. One reason for not returning the signal is that the tender should be able to see the tether line changing. More importantly, a 4 + 4 given in return could be confused with 4 + 4 + 4, a distress signal. Whatever the decision, it should be consistent throughout the team.

Rather than using left and right signals, some teams opt to use a single pull as meaning "Change direction." Simply changing direction, however, won't allow the diver to reorient himself to the tether, and it can soon foster confusion. When disorientation occurs, the diver can't help but focus on regaining his bearings, thereby decreasing his effectiveness in the search.

A "Change direction" signal also mandates the use of a "Take up the slack" signal. Without stopping to face the line, the tether often becomes slack. Worse, if the line is slack, the diver cannot feel the "Take up the slack" signal, so the tender actually has to take it in, destroying the search pattern. Also, without left and right signals, divers cannot be stopped for standby then restarted again in the same direction. Such divers have a less of a sense as to where objects lie within the search pattern, so their comments to profilers later are of decreased value in creating accurate profile maps. Opting to use a universal "Change direction" signal is a mistake that simply causes more mistakes.

Diver-to-Tender Signals

One pull: I'm okay. With this signal, the diver is simply letting all on shore know that he is okay, or that he is responding to a one-pull signal from the tender. The one-pull signal from the diver can also be used as a "Please repeat" by the diver.

Two pulls: Tender, make a notation. Two pulls mean that the diver would like a specific area noted on the profile map for later discussion and possibly another search.

Six plus six pulls: I found the target. This signal indicates that the diver has found the object of the search and is awaiting the signal to surface. As mentioned earlier, the 4 + 4 signal to surface may not be given immediately, pending the status of family members and others at the dive site. If the diver doesn't plan to surface with the object, as in the case of evidence at a crime scene, he must mark its location with a buoy. The lengthiness of this signal helps give a diver a chance to normalize his breathing before making the ascent.

Two plus two plus two pulls: I'm okay, but I'm tangled. Alert the backup diver. Since the backup diver is fully dressed, he should snap the contingency line onto the primary's tether if the tender receives a 2 + 2 + 2 signal. At the same time, the ninety-percent-ready diver should complete gearing up. Still, the backup shouldn't deploy unless the primary requests it. If ninety seconds pass after the original 2 + 2 + 2 signal without any further communication, then the backup diver should enter the water to assist the primary.

Three plus three plus three pulls: I'm okay, but I need help from the backup diver. The primary diver isn't dying, but he cannot get out of his situation alone. In the most likely scenario, the primary can't reach an entanglement to cut himself free. On receipt of this signal, the backup should deploy, but he can move slowly down the tether knowing that it isn't a life-and-death situation. The ninety-percent-ready diver prepares fully.

Four plus four plus four pulls, continuously: I need help immediately. In the case of an emergency, the diver should continuously give four pulls to indicate that he is still alive. The backup should deploy as quickly as possible, and the ninety-percent-ready diver should prepare fully.

Aside from mastering left and right, tenders may wish to rely on laminated cards showing all of the various signals. These memory aids can be clipped to a PFD for quick reference.

Line pulls are best given with a quick flick of the wrist. Long pulls can be mistaken for pulling the diver in, and giant pulls can pull a diver off track. If a tender has to pull hard to elicit a response, then there may be slack in the line, the line may be snagged, the diver's harness may be loose, or the diver may not be paying attention. In any case, the problem should be noted and corrected. Hard pulls can also cause accidental ascents, increasing the risks of entanglement and injury.

Sometimes divers fail to return a tender's signal because they're intent on the search. Sometimes it's because they're maneuvering around an obstacle and haven't had a chance to respond. Still, tenders should follow the three-strikes-you're-out rule: If a diver fails three times to respond to a given signal, send in the backup diver.

Surface Signals

While on the surface, tenders and divers can communicate using hand signals, although line signals may also be used. At night, divers will have trouble seeing hand signals. Tenders shouldn't arbitrarily shine lights on the divers, since doing so will ruin their night vision and blind them. While in the water, divers shouldn't communicate verbally unless they're using a communications system, since they'll have to remove their regulator to do so.

On surfacing, a diver should inflate his BCD and indicate that he's okay by touching one hand to the top of this head—unless, of course, he's unable to do so because he's holding on to a recovered object. If a diver surfaces and fails to give an "Okay" signal, the backup diver should deploy.

A diver can indicate that he's okay by touching one hand to the top of his head. (*Photo courtesy of Frederick E. Curtis.*)

If the tender wants the diver to move to a new location, he should point in the direction he wants the diver to move. Before divers swim out, and after they return, the tenders should signal, "What's your tank pressure?" This signal consists of tapping the open palm of one hand with two fingers of the other hand. The diver should report the pressure with one hand, each finger representing 100 psi. Air checks should be performed and recorded, along with the time, whenever a diver surfaces, even if he does so midway through a dive.

Underwater Contingency Signals

Primary divers and their backups must have a way of communicating underwater. Otherwise, the backup may not immediately be able to tell what the problem is. Having a rehearsed plan to follow will help lower the anxiety of the moment, better allowing the divers to solve the problem, get out, and go home.

After signaling for the backup diver, the primary should stop and put one hand on his harness carabiner. His only job at the moment is to concentrate on relaxing and slowing his breathing rate with long, slow exhalations. The primary diver will be able to feel the backup coming down the tether and should be confident that help is on the way. On arrival at the bottom, the backup will know where to find the primary diver's hand, since it is on the carabiner. The two divers should clasp hands so that the primary can indicate to the backup exactly what is wrong.

The divers use one or more of the following hand signals to communicate their needs.

Tapping the backup diver's hand to the primary diver's second stage. This signal means "I'm already on my pony bottle and I need more air." Air is always the number one priority for any underwater emergency. If the diver's main tank has been exhausted, this signal should be given immediately. Once this signal is given, the backup diver puts the primary diver's hand on the first stage of his (the backup's) pony bottle. The primary pulls the bottle and puts the mouthpiece in his mouth. The question is often asked, "Why not put the diver's hand on the pony mouthpiece?" The reason is that second stages can often become disconnected from where they were secured on the backup's BCD, so the primary might be unable to find it. The first stage, however, is always in the same place, and the hose of the first stage will always lead to the second stage.

The backup diver places the primary's hand back on the carabiner and gives three squeezes. This signal means, "I am leaving, but I'm coming right back." This sig-

The primary diver signals that he's out of air by tapping the backup diver's hand to his regulator.

The backup diver gives the primary his pony bottle.

nal should immediately follow the handing over of the pony bottle. Air is still the immediate concern, so the backup needs to get another pony, as well as a full tank for the primary. In returning, the backup should clip the contingency tank to the tether line so that it can't be lost, but he should handle its weight on the way down.

Large circular motion. This signal is used to show the backup diver where the entanglement is. The primary diver takes the backup's hand, traces a large circle with it, then places that hand on the problem area. Locating it for the backup this way helps him avoid getting snared on the problem himself.

Tapping the backup's hand on the primary's chest. This signal is used in the event of injury. The primary taps the backup's hand on his chest, then guides that hand to the affected area. The backup diver must move slowly and gently to prevent further injury.

Study Questions

1. Why are wireless communication systems prone to background noise?

2. When using an underwater communications system, why should a tender tell a diver to "Go three" or "Go four," rather than "Go left" or "Go right"?

3. What does a tender-to-diver signal of one pull mean?

4. What does a tender-to-diver signal of two plus two pulls mean?

5. What does a tender-to-diver signal of four plus four pulls mean?

6. What does a diver-to-tender signal of one pull mean?

7. What does a diver-to-tender signal of six plus six pulls mean?

8. What does it mean when a diver issues a continuous series of four pulls?

9. Immediately on surfacing, how should a diver indicate that he's okay?

10. After signaling for the backup diver, what should the primary diver do?

11. Once the backup diver makes contact with the primary, what should the two divers immediately do with their hands?

12. What does it mean if the primary taps the backup diver's hand to the primary's second stage?

Chapter 13

Contingency Procedures and Scene Safety

During the course of a search, it isn't uncommon for the primary diver's tether to become snagged on a rock, submerged log, or some other object. If a snag is detected early, the tender or diver can often solve the problem by dancing the line over the object. If this doesn't work, the tender can move in the direction that of the diver, thereby decreasing the angle of the tether around the snag enough to send a stop signal to the diver. With the primary stopped, the tender can either work to cure the problem, or he can send the diver in the opposite direction. If successful in unsnagging the line, the tender should move back to his original position to preserve the integrity of the search pattern. The profiler should note the location of the entanglement on the profile map. If many such snags occur, the team should either change the type of search pattern being used or split the area into halves on either side of the obstruction. If the primary diver is unable to clear the entanglement, then the backup diver should be sent in.

The first step for a backup diver in deploying to the bottom is to clip on to the primary diver's tether with a contingency line. Without a contingency line, the backup would be forced to hold on to or wrap his arm around the tether, which is a bad plan, since the backup could easily lose hold of it while equalizing or adjusting air in his BCD. The line floats, so once dropped, the backup would be forced to return to the surface to find it. Pulling hard on the tether could also further entangle or injure the primary diver.

If the backup comes to the snag along the tether, he should attempt to free it. If

Using a contingency line, the backup diver descends to the primary along the primary's tether.

necessary, he should disconnect from the contingency line, go past the object, and reconnect on the other side. On reaching the bottom, the backup assists the primary as needed. If the tether was too entangled to be freed, the backup then connects the contingency strap directly into the primary's harness carabiner and disconnects the primary's tether line. The two divers then make a direct ascent to the surface together, both tethered by the backup diver's line. Without the strap, the primary is no longer tethered once the tether line has been cut, and no backup can guarantee that he won't lose hold of the primary during the extrication, ascent, and surface swim.

Despite all precautions, there is a slim possibility that the diver may become disconnected from the tether. If that occurs, the diver must immediately initiate a safe ascent to the surface.

As mentioned elsewhere, there are very, very few circumstances in which a diver would intentionally disconnect from his tether. Even if he realizes that the tether is hopelessly entangled and line-pull signals are impossible, he should wait four to five minutes to see whether the surface crew has deployed a backup to retrieve him. If no backup arrives, the primary should cut himself free and ascend. On reaching the surface, he should inflate his BCD and signal his location to shore personnel.

If a disconnected diver is entangled on the bottom, he should try to relax and slow his breathing rate. His tender should realize that there is a problem, because the tether will go slack and there won't be any response to line pulls.

On discovery that a primary is disconnected, the backup should be sent out beyond the primary's last known location. Note the importance of marked lines and profiles for this type of rescue. The backup then descends. The tender then brings the backup toward the primary with short sweeps. He shouldn't try to bring the backup directly into the primary diver's bubbles, since the bubbles may not be directly overhead the primary. Also, if the backup misses his target, precious time will likely be wasted with shotgunning attempts. Once the two sets of bubbles are in the same place, the divers should be together.

If the disconnected diver has stopped breathing or if the tender can't see his bubbles, the backup diver should be sent fifteen to twenty feet beyond the primary's last known location and begin running a small search pattern within the likely area. The backup diver's tether will probably snag on the downed primary even before the backup has found him. The backup diver should recognize the resistance he feels on the line and go straight to that location.

Rescuing an Unconscious Diver

If the backup diver approaches the primary and finds him lying still, the backup should gently tap the primary to check for his responsiveness. If the primary's full-face mask is still on or if the regulator is still in his mouth, the backup should leave them in place. If either has been removed, the backup should make no attempt to replace them. The immediate task is to bring the primary to the surface. Unconscious divers can't hold their breath, so lung overexpansion injury is unlikely to be a problem, and in this case, decompression sickness is not the predominant concern. If the unconscious diver is wearing weights, the easiest way to get him to the surface is to release those weights. The backup diver should not continue to hold the primary once he begins to ascend faster than sixty feet per minute. Rather, he should let the primary float to the surface, where the ninety-percent-ready diver should be ready to retrieve him.

If the diver is face-down and unconscious at the surface, reach out with your right hand and gently push him on his right shoulder as you shout to check for his responsiveness. As his body floats backward a foot or so, his arms will most likely float upward. Grab his right wrist with your right hand, and bring the wrist downward in a big arc to the right to rotate him into a face-up position. Otherwise, you may roll the diver over by pulling his tank down to one side.

Once the primary is on his back, signal the tenders to pull you both in gently as you pass your left hand under his left armpit and grab his tank valve. The left-hand position is the most effective because it puts the rescuer on the diver's left side, near the power inflator, and it allows the rescuer to use his right hand to release the weight belt. If the primary's weight belt is still present, reach over his body with your right hand and release the belt while simultaneously pushing down on his stomach to angle his torso downward. This will prevent the belt from landing on the tank. If the diver tightened the belt at depth because of suit compression, the belt will be too tight back at the surface when the suit reexpands. Therefore, just removing the belt might be enough to get the diver to start breathing again.

In one smooth motion, move up the diver's body from the waist, releasing his BCD cummerbund, chest strap, dry-suit inflator, and anything else crossing his torso. This step has three functions. First, the primary's breathing may have been restricted by the BCD straps having been pulled too tightly. Second, it reduces the diver's gear, thus lessening the amount of drag involved in pulling the primary to shore. Third, it'll allow you to bring him into a boat or on shore more easily.

If the water is known not to be contaminated, continue working your way up the diver's body. Remove the regulator with an upward motion to open the airway, then remove the mask. This could also possibly stimulate the diver to start breathing.

If the primary is tethered, tenders should slowly pull him in, along with the backup, at a rate of no more than one foot per second. Any faster might cause the primary's head to submerge.

The backup should protect the victim's airway from water by coddling him. To coddle a victim, place your right arm under the neck of the diver, and bring your hand up over the diver's mouth and nose to block water from entering his airway.

Tethered backups shouldn't administer in-water rescue breathing unless the rescuer is well trained and practiced in this technique, and also if it won't mean exposure to contaminated water. Since there's a chance that the unconscious diver still has a heartbeat, in-water ventilations may be of benefit. However, tests with mannequins have shown that significant amounts of water can be blown into a victim's lungs by this procedure. If the divers aren't tethered, in-water rescue breathing will delay reaching a stable platform. Moreover, since cardiopulmonary resuscitation can't be administered in the water, attempts at ventilation will lengthen the time that the victim goes without compressions, if needed.

Make sure that the team has a practiced plan as to how the diver will be extricated from the water and transported to the ambulance.

Crowd Control

Just as with any other emergency situation, crowd control at a dive site is necessary for both preventing bystanders from interfering with an operation and for keeping everyone safe. Would-be freelance rescuers, curiosity seekers, reporters, and others at the site should be kept where they cannot damage equipment, interfere with communication, destroy evidence, or simply get in the way. Under no circumstances should bystanders be allowed beyond the cold zone. In fact, even fire, police, and EMS personnel not trained in diving operations should be kept in the cold zone, if possible. Although they may have the best of intentions, untrained personnel may easily hamper an operation. A good compromise in using such personnel to good effect is to assign them crowd-control duties.

Crowd control should start with the arrival of the first unit, and a distinct group or sector should be given the task as the incident expands. The best way to secure an area from a crowd is to define the perimeters of the various zones, whether by tape, rope, flares, barricades, fences, vehicles, or some other means. Always secure an area larger than the area you need. The primary fault in crowd safety usually lies in the failure to establish a large enough zone. It's always easier to establish a larger area than needed at the outset and to reduce it than it is to push back a crowd or rearrange the operations later.

To remove uncooperative people from restricted areas quickly and easily, ask for identification, since they are within a possible crime scene. Also, suggest that it is in their best interest to provide identification, since they may have touched something and their fingerprints will identify them. Ask for names and contact information, and explain that the list will be used as documentation that they were present and watching. If they leave without saying anything, you've accomplished your task. Most people fear being dragged into court to testify and would rather skedaddle than push the issue.

Family Members

One of the most important aspects of scene safety is how the team deals with the relatives of a victim lost in the water. The key to working with a victim's family is to treat them as you would want to be treated. Talk to the family at each step of the operation, and be totally honest with them. Choose a team member to act in a liaison capacity, keeping them abreast of the rescue effort. The appearance of professionalism is critical in dealing with a victim's family. Anything less will give the appearance of incompetence. Disenchanted families can be uncooperative and critical, which, in front of the media, can be extremely damaging to a team's reputation and status.

As a member of a dive team, you may sometimes experience difficulty in dealing with family members. A relative may lash out at you, angry that you haven't yet brought the victim back to them. Sometimes grief can overcome a person so much that he will deny that a family member is dead, believing instead that the loved one is "just missing in the water." All of these responses are normal. Personnel shouldn't take errant emotions personally. Rather, they should constantly remind themselves that grief can make people act irrationally.

Despite the importance of communication and honesty, family members shouldn't be allowed beyond the cold zone, both for their own safety and that of the operation. If possible, keep family members in an official vehicle to shelter them from others, as well as to help them feel a part of the operation. Be sure, however, that the radio in the vehicle is turned off.

Diver-Down Flags

Use diver-down flags to facilitate another form of crowd control: that of restricting boat traffic around the dive site. Two different flags are used to indicate that there are divers in the water. The first, red with a diagonal white stripe, is commonly used in the United States. Secured to a diver's float, buoy, or raft, it means that there is a diver in the water and that vessels should keep clear. The second is the international code flag Alpha. This flag is technically flown for vessels that are restricted in their movement, thus signaling other vessels to keep clear, but it has also come to mean "Diver down, keep clear." The international-water rules for operating vessels specifically require a boat, when engaged in diving operations, to display a rigid replica of the Alpha flag. Exhibiting this signal invokes special right-of-way privileges that this boat has over essentially all other vessels. Because of this, the Alpha flag should only be flown on a vessel, not on a diver's float or raft.

Either or both of these flags, as appropriate, should be flown for any operation in an area that may have boat traffic. This holds true even for land-based operations,

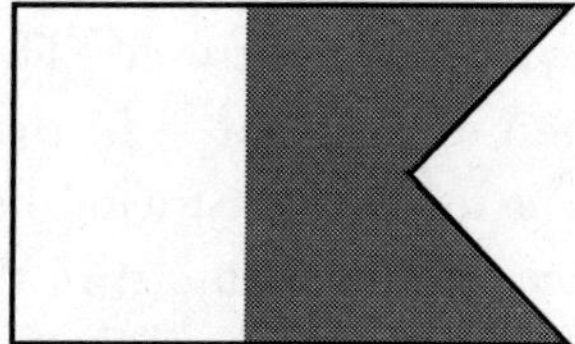

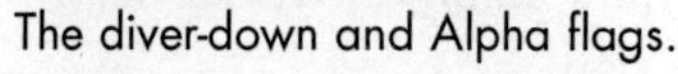

The diver-down and Alpha flags.

If necessary, call in law enforcement boats to help secure the search area.

since divers may be deployed to as much as nearly 150 feet from shore. It may be necessary to use buoys or boats to fly flags to demarcate the perimeter of the search area.

Laws as to how far boats must keep from dive flags vary from state to state, as well as internationally. The distances range from 50 to 150 feet. Divers are often required to surface within 25 feet of the flag. Obviously, boat traffic through the search area would be extremely hazardous to the divers, so be sure to mark the boundaries with suitable flags. The flags should be no more than 100 feet apart along the perimeter of the operation to ensure that boaters see and obey them. Also, if you are conducting diving operations at night, any vessels must be clearly marked with navigation lights.

Perhaps most importantly, remember that flags aren't magical boat repellents. Some people are ignorant about their meaning, some won't see them, and some won't care. Be particularly alert to private boats, whose pilots tend to be novices and may be moving too fast to take notice, anyway. If necessary, call in law enforcement boats to help secure the search area.

Rehabilitation

The incident commander should establish an rehabilitation area whenever an incident will last more than forty minutes; when the first diver has exited the water and the incident is likely to continue; when the number of injuries exceeds three; and when the ambient temperature is either significantly hot or cold. As mentioned earlier, this area should be set up away from the incident, and it should provide protection from the elements. Once the site has been chosen, it should be staffed by medically trained personnel who have knowledge of scuba-related injuries. Personnel assigned to the rehab should check the vital signs of all incoming personnel, noting whether any diver shows signs of decompression illness. The divers should drink non-caffeinated fluids, and they can only return to diving if they have been medically cleared. Under no circumstances should a diver who shows signs of dehydration be permitted to dive again.

For hot environments, the supplies that should be available at the rehab site include water, ice, juice, fruit, and energy bars, but no sugary snacks. Some means of keeping the divers cool, whether by fans, air-conditioning, or shade, is important. Towels and first aid supplies are standard. For cold environments, rehab should be stocked with warm water, broth, and other warm fluids, as well as fruit, soup, and energy bars. Towels, heaters, blankets, spare gloves, hats, dry clothing, and first aid supplies should all be available. A heat-retaining recovery suit may prove to be a lifesaver. Items that should not be found there include sugary snacks, caffeine, and sports drinks.

Lengthy Operations

If your team is searching a large area, the weather is turning bad, personnel are exhausted, or the rescue becomes a recovery, you may be faced with an operation that will stretch over a number of days. In fact, if circumstances warrant, you shouldn't hesitate to shut down a recovery operation and return later to continue the search. Spreading a difficult recovery over several days can actually be beneficial, since it will allow personnel recuperate while maintenance is performed on the equipment.

In deciding to spread a potentially difficult operation over several days, determine the area you have to search and the approximate amount of time it may take you to cover it. That way, you can plan for personnel, food, water, equipment, and other resources in advance.

If you opt to shut down an operation temporarily, don't leave the scene unsecured. Leave any police tape or other demarcations in place to keep curiosity seekers from the warm zone. If possible, request assistance from local law enforcement. If the police can't post an officer and a patrol car at the site, ask them to patrol it regularly while the team is gone.

Such measures for securing will not only discourage people from tampering with the site, they'll also serve to reassure the victim's family and the community that the team hasn't simply given up and gone home. To help ward off wrong impressions, you may also wish to inform the media of your decision to suspend operations and explain why you have chosen to do so. Otherwise, reporters may speculate that your team is simply stumped.

Helicopters

The decision to call for a helicopter is made by the incident commander. Transport by helicopter should be strongly considered in cases of unconsciousness, uncontrolled bleeding, multiple fractures, severe shock, hypothermia, amputation, spinal injury, or decompression illness. Helicopters should be used when ground transportation to the appropriate care facility will exceed thirty minutes, or whenever other circumstances of resources and logistics demand. If a helicopter is used for the transportation of a diver, be sure that the crew is aware of altitude concerns vis a vis recent diving and suspected decompression illness.

Medical helicopters usually won't respond for patients in cardiac arrest. Often there is one exception, of which responding ambulance personnel may not even be aware. That exception is for victims in cardiac arrest due to long-term drowning and the vic-

A helicopter cannot respond unless there is an appropriate landing site.

tim was recovered in a rescue mode. Check your local protocols and spread the information to the appropriate personnel.

When calling for a helicopter, you'll probably be asked for some specific information, so have it written down and ready. Be prepared to give sound answers to questions about your exact location, landing zones, the number of victims, the nature of the injuries, weather, wind direction, and any special equipment required.

If the helicopter is being used to transport a diver, be sure that the crew is aware of altitude concerns with respect to flying after diving and flying with suspected decompression illness.

A helicopter cannot respond unless there is an appropriate landing site. If there isn't one available nearby, you may have to transport the patient by ground to meet the helicopter elsewhere. The landing zone should be flat, firm, and level. It must be free of high obstructions, such as telephone poles, wires, streetlights, and trees. The zone should have clearance of at least fifty feet on all sides. Besides obstructions, this fifty-foot clearance also applies to vehicles, personnel, and lightweight material that might be blown about by the rotorwash. If dust appears to be a problem, wet down the area. Control all animals and bystanders. Stop all vehicles. Spectators should be restricted to at least two hundred feet from the touchdown point. For night operations, set out lights or flares to designate the landing zone, but ensure that any lights point downward so that they won't blind the pilot as he makes his approach. Never activate smoke canisters upwind of the landing site.

Some basic safety tips apply to working around helicopters. Anytime the engine is running, request the pilot's permission before going near, and always avoid the tail. Instead, approach the craft from the front, preferably biased toward the pilot's side (starboard), so that you can see his eyes. Crouch as you get near, and never carry long items vertically. Never wear a hat. If the helicopter is parked on a slope, approach it from the downhill side, never from above. If a hoist is being used on a hovering helicopter, allow the hoist to contact the ground before you touch it. This will allow static electricity to discharge.

Study Questions

1. What is the first step for a backup diver in deploying to the bottom?

2. How does the backup diver prepare the primary diver for the ascent?

3. On discovery that the primary is disconnected, how should the tender direct the backup diver to the primary diver's location?

4. If the backup diver finds that an unconscious primary has lost his regulator or mask, should he try to replace it?

5. If a backup diver releases the weights from an unconscious primary diver to facilitate an immediate ascent, why won't lung overexpansion injury be a predominant concern?

6. Once the primary diver is at the surface, why shouldn't the tenders pull him to the boat faster than one foot per second?

7. What is perhaps the surest way to remove uncooperative people from restricted areas quickly and easily?

8. The international rules for operating vessels specifically require a boat, when engaged in diving operations, to display a rigid replica of the ________.

9. Why should you inform the media if the team decides to suspend operations?

10. A medevac helicopter should be called in when ground transportation to the appropriate care facility will exceed ________ minutes.

11. The landing zone for a medevac helicopter should have clearance of at least ________ feet on all sides.

12. If a hoist is being used on a hovering helicopter, why must you allow the hoist to contact the ground first before you touch it?

CHAPTER 14

Physiological Components of Diving

The layman's definition of hypothermia might be "the body's inability to maintain normal core temperature" or "a loss of core heat." The medical definition, by comparison, is a body core temperature of less than 95°F, or more than three degrees below normal. The difference between these two definitions can present a problem for rescuer safety. A person using the clinical standard might infer that anything short of a three-degree loss isn't a serious problem, since it doesn't constitute hypothermia. This is far from correct, however. How a person feels can be just as important, if not more so, than the actual quantity of heat lost.

The effects of cold and feeling cold can result in physical, mental, and emotional aberrations well before the onset of clinical hypothermia. This book, therefore, will operationally define cold stress in sum as the direct or indirect effects of heat loss, irrespective of any specific drop of core temperature. Those effects can include poor judgment, confusion, fear, panic, hallucinations, irritability, fatigue, indexterity, loss of strength, tingling, numbness, pain, and out-of-character behavior.

Cold stress and especially hypothermia carry with them more than simple discomfort. Both can be deceptive killers of public safety divers. One of the most insidious dangers of cold stress is that, in his desire to return to the surface, a diver may be lured into taking shortcuts. Such shortcutting can be the start of the accident chain, in which small problems compound into greater and greater ones, perhaps leading up to an injury or fatality. Even a drop in skin temperature will result in a release of the hormone epinephrine, which is part of the fight-or-flight mechanism.

That release adds to a release already caused by the incident itself, and it can lead to fear, panic, mental disorganization, and a loss of judgment. The problem isn't limited to divers, for all personnel at a rescue site on a chilly day may be at risk. Critical members such as tenders should particularly beware that they don't unwittingly stand at the beginning of an accident chain.

The Mechanics of Heat Loss

Water conducts heat away from an object twenty-five times faster than air does, meaning that body heat will be stolen from you far faster in water than in air. In terms of heat capacity, it takes more than 3,200 times more heat to raise a given quantity of water one degree than it does to raise an equivalent parcel of air the same amount. Because your body is in direct contact with the water, you will continue to lose heat until you and the water are the same temperature. Unless you think that you can readily heat an entire lake, you need suitable exposure protection.

The conductive properties of water are so pronounced that, in most areas, some type of thermal protection is critical, even in the summer months, and especially for repetitive dives. Although your body will continue to produce heat in an effort to maintain your core and skin temperatures, water of anything less than nearly body temperature will continue to steal it away. The human body loses heat at the same rate in 80°F water as it does in 42°F air. Core temperature is lost in water of anything less than about 92°F, which is about the average skin temperature. Every time that you are immersed, you're losing heat, even in water that feels warm. That heat loss presents a good reason not to put backup divers in the water until they're needed. Otherwise, you may have to rotate a cold, wet backup into rehab rather than into the position of primary diver.

The effects of evaporation are also of concern to diving operations. As it evaporates, the water clinging to a diver's skin and exposure suit will steal heat from his body, using that heat in converting to a vapor. Evaporation from the skin and respiratory tract normally account for about 20 percent of the heat lost from the body. If the skin is wet from an external cause, as from swimming or the rain, that percentage increases. Because the process of evaporation can steal so much heat, it is important for divers to dry off as quickly as possible after exiting the water, particularly during cold weather. Still, it's common for a diver to leave his wet suit on during the surface interval. Although he may think that he's warmer with the suit on, he's being deceived by his own body. The problem is that our perception of warmness or coldness is based on skin temperature, not the relative warmth of our core or the amount of heat we're losing. Thus, even if a wet suit makes a diver feel warm, he may actually be losing heat.

Even during the summer, leaving a wet suit on can chill a diver to the point that he'll be unable to perform a second dive. Water trapped within the suit will continue to sap heat away from his body, while evaporation on the outside will deprive him of still more. The heat loss will be further hastened by the wind, which will speed up the evaporation process and introduce the effects of convection, related to the movement of fluids. Unless the air is warm and still, there will simply be no way for a diver in wet gear to warm up, and any movement of cold water against his skin when he is immersed will convectively steal heat away at a prodigious rate.

Through the act of breathing, too, the human body loses heat through both convection and evaporation. When a person takes a breath, the airway and lungs moisturize and heat the inhaled air to body temperature. During exhalation, that warm, moist air leaves the body, and the new breath must be heated. The problem is worse for scuba divers. Since the air from a scuba tank is cold and dry, expanding out of its compressed state, a diver must expend even more heat and moisture in heating up each lungful. Although they may not realize it, breathing air from a scuba tank can instigate heat loss in any diver.

The lesson inherent to all of these statements is a straightforward one. Although many divers are fairly cavalier about the issue, the need to remain warm, dry, and hydrated during water operations cannot be overemphasized.

The Cold-Stress Curve

Heat loss often occurs at an increasing rate. For example, if it takes twenty minutes to lose one degree of core temperature, it might only take another ten minutes to lose a second degree, and an additional five minutes to lose a third. Once hypothermic, it could take as much as twenty-four hours to rewarm the body to the same physiological state you were in before any heat loss occurred. If you don't wait a full twenty-four hours to rewarm thoroughly before going back outdoors, you'll become chilled even faster than you did the first time.

The cold-stress curve explains why divers gradually get more and more chilled over the course of an operation. More importantly, it shows why divers cannot be used for numerous dives in cold water or in cold weather, even though they may be well within decompression limits. Each minute spent in the water or cold air makes it increasingly harder to warm up again and be prepared for the next dive.

Naturally, the cold-stress curve as shown is descriptive. Its true magnitudes differ among individuals and are probably never constant even for the same person. People with little or no body fat tend to lose heat faster than those with more body fat, and they'll usually feel colder faster. Women will typically feel colder faster because their peripheral blood vessels constrict earlier, shunting more blood to the core, where it remains relatively warm. Women may actually lose less heat than men; however, their generally lower metabolisms also mean that they may produce less heat than men and so can less afford a deficit.

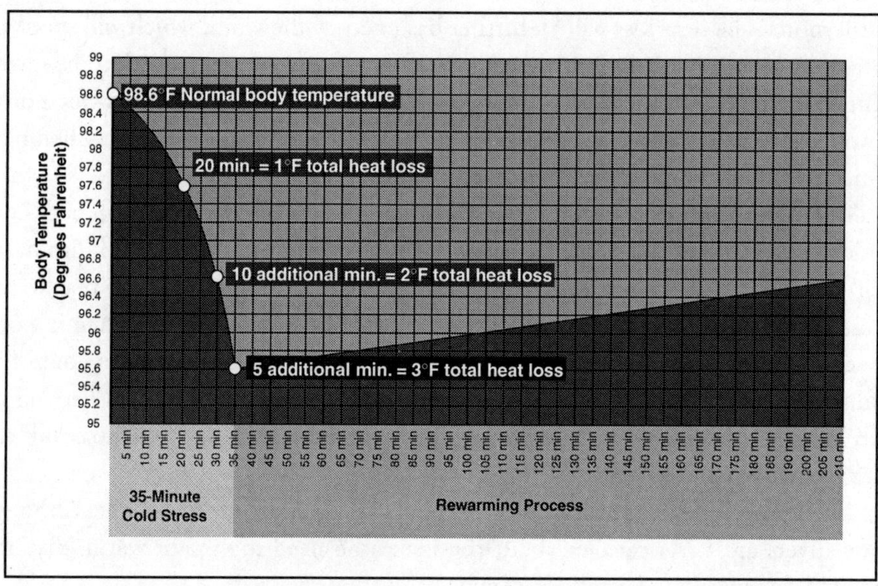

An example of a cold-stress curve.

Denial

Cold stress decreases your ability to function safety, but it does so subtly and gradually. Because it creeps up without warning, a diver doesn't usually notice it until it's too late. Thought patterns, reaction time, and muscle coordination become poor. The most dangerous aspect of the condition is the denial that accompanies it. Instead of taking appropriate measures, the diver convinces himself that he can handle it "just a little while longer." On land, we can usually tell when the weather has gotten to us and it's time to go inside. In the alien realm of the water, however, our sense of self-preservation isn't nearly so well attuned.

When it comes to safety and exposure, we shouldn't give in to the natural urge to take warm-weather dive operations any less seriously than we do cold-weather operations. In fact, there may even be a greater incidence of cold stress and hypothermia during the warm months, when many divers abstain from using such items as a hood, gloves, or a proper exposure suit. The threat of hypothermia doesn't disappear just because the summer sun is out or because you happen to live in a warm climate. The danger is there all the time.

Unlike most divers, public safety divers must exert themselves underwater. Rescue and recovery operations entail working dives, and so require more physical effort. Unfortunately, exertion means increased heat loss, even though the diver may feel warmer.

As mentioned above, we judge how warm or cold we feel largely by the temperature of our skin. We feel warmer during periods of physical exertion because our peripheral blood vessels dilate, thereby increasing the flow of blood near the skin and in the limbs. Even though we may feel warmer, more blood is ferrying more body heat from the core to the limbs and skin, where water steals that heat away. Divers who are exerting themselves may not even realize the amount of body heat that they're losing—until, of course, they slow down or approach the end of their dive. Then they'll feel absolutely exhausted and chilled.

Conduction robs heat from the body in areas where there is little insulation and where blood vessels, carrying blood heated in the body's core, run close to the surface. About fifteen percent of the body's heat is lost through the hands, and another fifteen percent is lost through the feet. When the hands are left unprotected, the muscles become paralytic. The head accounts for only about seven to nine percent of the total surface area of the body, but we lose about twenty-five percent of our body heat from the head alone. Because of that tremendous heat loss, a diver should wear a hood even if he thinks he can "get away without one." At the start of an operation, a diver doesn't know how many dives he'll have to perform, so he can't afford to get cold at the beginning.

Coldness also has implications for decompression illness. Many of the symptoms listed above for cold stress and hypothermia are also common to decompression sickness and arterial gas embolism. This similarity of symptoms may potentially prove problematic in determining the root cause of a condition. Suppose, for example, that on an 80°F day in April, you rushed to get in the water to perform a possible rescue, and in doing so, you neglected to wear a hood and gloves with your wet suit. You made a dive for twenty-five minutes to forty-five feet in 60°F water, then accidentally ascended too rapidly in the black water. When you surfaced, you were shivering beyond control, and your face, hands, and feet were numb and tingling. You stayed in your wet suit in case you might be needed to dive again. Now, an hour after surfacing, you're feeling better, but your left hand is still tingly and numb. What is it? Are the sensations left over from the cold, or are you suffering from an arterial gas embolism? The diver who is in doubt should always follow the protocols for decompression illness, the high end of which can include oxygen administration and treatment in a hyperbaric facility.

In point of fact, cold is believed to be a predisposing factor toward decompression sickness. Descents made in colder water and those that are arduous should both be treated as longer dives when using dive tables.

Accidental Immersion

If you fall into the water while not wearing an exposure suit, your PFD will at least keep you afloat. If the water is cold, however, you'll find yourself in danger of hypothermia due to the effects of conduction, convection, and evaporation. Unless you can easily get out again, do not tread water and kick around furiously. Instead, try to conserve energy by huddling. To huddle simply means to pull your legs up and hold them to your chest with your arms. That position reduces the surface area of the body exposed to water, thus increasing your ability to retain heat. Although a person will feel colder, huddling will actually result in a reduced loss of heat and a greater chance of survival.

Even if you aren't wearing a PFD, you may be buoyant enough to huddle. If not, use the bobbing technique. Bobbing involves floating vertically in the water, your head tilted back and your face exposed. Take a breath and allow yourself to submerge, without using any energy to stay afloat. Then, gently exhale your breath underwater, and as your lungs empty of air, use your arms and legs in small motions to propel yourself slowly back to the surface. Roll your head back, and again, with only your face exposed, take a new breath. The keys to bobbing are to stay relaxed and to use as little effort as possible. The less you move and the less you raise yourself above the surface, the less energy you'll use. Because you're using less muscle, more heat will remain in the core of your body, and because moving less will decrease the amount of water flowing past your skin, the less heat you'll lose due to convection.

HYPERTHERMIA

No discussion of the risks of cold stress and hypothermia could be quite complete without mention of their opposite, hyperthermia. Although water can chill a diver, donning and wearing an exposure suit on land during warm weather can be a miserable experience. Sometimes a diver can get so hot in gearing up that he forgets, intentionally or otherwise, to put on the thermal protection he'll need underwater. A backup diver waiting onshore in the sun is especially prone to overheating. The real problem is that overheating can be more than uncomfortable. It can be dangerous.

While suiting up, a diver can easily become overheated, raising his blood pressure enough to prevent him from diving. Overheating can eventually cause heat cramps, heat exhaustion, or heat stroke.

Heat cramps, usually in the legs or abdomen, occur when heavy perspiration deprives the body of salts. That salt deprivation means that the muscles no longer have what they need to function properly, so they start cramping. If team members experience heat cramps, they should be moved to a cool place away from the water where they can take off any gear that they're wearing. Massage the cramped muscles, and provide fluids replete with electrolytes. Apply moist towels to cool the patient. If the cramps persist, seek further medical aid.

Heat exhaustion is a condition caused by dehydration and the loss of body salts. It generally occurs in individuals working hard in hot conditions. Its symptoms are rapid, shallow breathing; weak pulse; profuse sweating; cool, clammy skin; dizziness and weakness; and nausea. To treat it, loosen or remove any tight gear that the patient might be wearing. Move the victim to a shady area and administer fluids, which should be ingested in sips. Don't try to cool the body too quickly. Rapid cooling will cause shivering, which will only elevate body temperature. As a final treatment, administer high-flow oxygen.

Heat stroke is a dangerous condition in which the body's heat-regulating mechanisms fail. Its most critical symptom is any change in mental status or in functions related to the central nervous system. The breathing will be deep at first, then shallow. Similarly, the pulse at first will be rapid and pounding, then weak, although still rapid. The skin will be dry and hot, and the pupils will be dilated. Loss of consciousness may occur. In treating heat stroke, cool the patient in any manner and as rapidly as possible. If possible, place wrapped ice packs under each armpit, behind the knees, on the groin, the wrists and ankles, and on either side of the neck. Treat the patient for shock, administer high-flow oxygen, and transport him immediately.

You fend off overheating among dive team members in a number of ways, many of them fairly obvious. The prescription to drink plenty of fluids, avoid overexertion in the sun, and wear appropriately light clothing in the summer applies to all shore personnel, not just divers. One of the reasons a tender helps a diver gear up is specifically to keep that diver from overheating. Once suited up, backups and ninety-percent-ready divers can sit or hang in the water if the diving coordinator and safety officer allow it. You may also douse them with clean, fresh water, or cover them with a damp towel. Dry-suit divers can pull out their neck seals and press their suit inflator button for ten seconds, which will rapidly cool the body while expending a minimal amount of air.

A common mistake is to keep divers who are fully geared up standing for too long. Once a diver has finished his gear check, he should enter the water or sit somewhere near it. After the dive, have him remove all of his gear. Although many divers like to leave their exposure suit on, doing so may bring on overheating.

Blood Pressure and Volume

Even before a diver enters the water, his blood pressure can be raised by stress and his exertions in gearing up. It's when he enters the water, though, that the real trouble can begin. The spike in blood pressure that he experiences stems from two physiological changes that occur when a diver enters the pressure-laden, possibly cold, environment below the surface.

Water is eight hundred times denser than air, so it weighs much more and exerts far more pressure. The increase of water pressure for every foot of depth is approximately 0.44 psi. Thus, a six-foot-tall person hanging vertically in the water has approximately 2.6 psi more pressure exerted on his feet than on his head. As a result, a significant amount of blood is squeezed out of the extremities of the body, pooling instead in the body's core, causing the blood pressure to rise. The blood pressure also rises as an effect of the cold, since the blood vessels near the skin constrict, shunting blood to the core in an effort to keep the body warm. This increase in blood pressure is partly responsible for the universal urge to urinate when we are cold or submerged. The body tries to reduce blood pressure by decreasing blood volume through the production of urine. The colder the environment, the more vasoconstriction occurs, the more peripheral blood is sent to the core, the higher the blood pressure, the more the brain tells the kidneys to pull water from the blood to lower the blood pressure, the more urine is produced, and the lower the blood volume.

If a person begins with high blood pressure, cold water can raise that blood pressure to dangerous levels. Factor in the effects of exertion and dehydration, and this person may be in trouble.

Dehydration

The processes described above have another implication for diving, in that the increased production of urine can dehydrate the body. Dehydration presents several important concerns for all scuba divers, including fatigue and irritability; lowered blood pressure; muscle and heat cramps; increased risk of shock if an injury occurs; hypothermia; hyperthermia; and an increased risk of decompression sickness. All of these conditions mandate that underwater personnel pay particular attention to their fluid intake before the start of an operation, as well as afterward. Compressed air, sweating, caffeine, and alcohol can all serve to induce dehydration, and some public safety divers will even avoid drinking liquids before making a descent in a misguided effort to stave off the need to urinate while underwater.

Given all the ways in which a diver might become dehydrated, it's important that team members learn to recognize the symptoms. If you don't feel the urge to urinate when you're underwater, then you're most likely dehydrated. Dry mouth, thirst, and an inability to spit or sweat are late symptoms, as is dark urine. Irritability, fatigue, and headaches are danger signs that you might need more fluids. Dry eyes often indicate dehydration. Muscle cramping can be a symptom, and if you can pinch your skin and it doesn't immediately return to normal, you are very dehydrated.

To be sure you are well hydrated, make sure your urine is clear, copious, and frequent. The rule is simple: Drink plenty of water, both before and after dives, continuously throughout an operation. Water is the best fluid for divers. Although sports drinks do rehydrate the body, water loss, not electrolyte loss, is the main concern for divers.

Study Questions

1. What is the medical definition of hypothermia?

2. Water conducts heat away from an object how many times faster than air?

3. It takes more than ________ times more heat to raise a given quantity of water one degree than it does to raise an equivalent parcel of air one degree.

4. Our perception of warmness or coldness is generally based on ________.

5. Why will women typically feel cold sooner than men?

6. Why do we feel warmer during periods of physical exertion?

7. True or false: Dives made in colder water and those that are arduous should both be treated as longer dives when using dive tables.

8. Name two techniques for conserving energy in the water.

9. If a diver is suffering from heat exhaustion, why shouldn't you try to cool him off quickly?

10. What is the treatment for someone exhibiting the symptoms of heat stroke?

11. Dark urine, dry mouth, thirst, and an inability to spit or sweat are late symptoms of ________.

Chapter 15

Diving Maladies

A thirty-six-year-old police officer exits the water after making a search for a handgun. It is early March, and the water is only about 40°F. His dry suit leaked, and now he is miserably cold. In addition, he is feeling very tired and is complaining of some pain under his right shoulder blade. He did make too quick of an ascent after his suit leaked, but he thinks that the pain is due to lifting tanks out of the truck.

Since most EMS personnel aren't trained in dive-related maladies, the question for the safety officer becomes one of how to evaluate a diver who surfaced too quickly and is complaining of pain in his right shoulder.

Diving maladies are generally grouped together and described as decompression illness, or DCI. However, there are two general types of DCI; namely, lung overexpansion injury and decompression sickness (DCS). Although diving maladies should only be treated by qualified medical personnel, it's important that divers and safety officers understand what types of diving injuries might occur and what their causes may be.

Lung Overexpansion Injury

There are several different types of lung overexpansion injuries, including arterial gas embolism, pneumothorax, mediastinal emphysema, and subcutaneous emphysema. What they all have in common is that each can be induced by holding your breath for any reason during an ascent while on scuba or surface-supplied air.

How do you evaluate a diver who has surfaced too quickly and is complaining of pain in his right shoulder?

According to Boyle's law, the volume of a parcel of gas varies inversely with the pressure exerted on it. If the pressure increases twofold, the volume of the gas will be half its original volume, and its density will be twice as great. When a diver goes underwater, the pressure increases on his body due to the weight of the water above him. A skin diver with five liters of air in his lungs will still be holding the same amount of air at ninety-nine feet, where the pressure is four times what it is at sea level, but because pressure and volume are inversely related, that five liters will compress and theoretically only occupy one-fourth the space it occupied at the surface. As long as the skin diver doesn't inhale or exhale while he's underwater, his lungs will return to their normal size, and the air within them to its normal volume and density, when he returns to the surface.

A scuba diver is different from a skin diver, because a scuba diver breathes underwater. With each inhalation, a scuba diver fills his lungs with compressed air to keep them the same size that they were at the surface. This means that he must breathe increasingly denser air as he descends. To maintain a five-liter lung volume, a diver must inhale the surface equivalent of ten liters of compressed air at thirty-three feet, fifteen liters of compressed air at sixty-six feet, and twenty liters of compressed air at ninety-nine feet. This is the function of a regulator—to deliver air at ambient pressure according to the depth. If a scuba diver holds his breath while ascending, the denser air that he used at depth will expand the lungs beyond their normal elastic capability, causing tissue damage. A three- or four-foot ascent while holding your breath with your lungs full could be enough to cause perforations in the alveoli or other injury to the structures of the respiratory system.

Scuba divers are trained to breathe continuously and normally underwater, but there are many situations in which a diver may unwittingly, unconsciously fail to follow this rule. Some of those situations, and others leading to possible lung overexpansion injury, are as described below.

1. While equalizing. Divers should never perform breath-holding equalization methods such as swallowing or the Valsalva maneuver while ascending. Perform gentle Valsalva maneuvers without having your lungs filled with air, and make sure that your ascent has stopped. Better still, use an even gentler technique. Hold your nose, put your tongue on the roof of your mouth, and swallow. Although this technique still involves a breath hold, it doesn't have the back pressure of the Valsalva maneuver.

Divers should equalize early and often, first on the surface, and every one to three feet thereafter on the descent. Sometimes a variably sloping bottom will cause a diver to go up and down with every sweep of the search pattern. In black water, the diver may not even be aware that he is doing so. Still, such frequent depth changes increase the need to equalize, while also increasing the chances that the diver will accidentally equalize while ascending. Once on the bottom, a diver can avoid that problem by stopping each time he needs to equalize.

2. While dealing with mask-related problems. Watch other divers for signs that they may not be comfortable with their masks. Discomfort indicates the potential for a diver with a dislodged mask to panic, bolt, and be injured. If a diver continually rips the mask off his face every time he surfaces, it's a sign that he's not completely comfortable wearing it. Another sign of discomfort is a large, red ring on the face where the mask sits. If the mask strap isn't too tight, then that ring will mean that the diver has constantly been trying to inhale through his nose, thereby sucking the mask against his face. Nose breathing is an indication that a diver will tend to panic if his mask becomes dislodged.

Also, remember that you don't need to take a large inhalation to clear your mask. The volume of your mask is far smaller than that of your lungs, so you can clear it with only a partial lungful.

3. While coughing, sneezing, hiccuping, or vomiting. Any of these actions underwater are common occurrences, but they do require a little special attention. When people cough, sneeze, hiccup, or vomit, the first thing they do is inhale sharply, filling their lungs. Just as with equalizing, the risk is posed by holding a lungful of air during the ascent, and this risk is compounded because a sneeze or a cough also means a contraction of the body and sudden, forceful back pressure.

Arrest your ascent whenever you feel the need to cough, sneeze, hiccup, or vomit. Bear in mind that, as you reflexively fill your lungs, you will inadvertently tend to rise because your lungs will have been made more buoyant.

4. While working hard or concentrating. If a diver is working hard, then he is also breathing heavily, with full inhalations and exhalations. Such a diver may be prone to hyperbreathing, which is a spiraling process. The diver starts out breathing normally at the beginning of a dive. As he continues to work hard, he quickly begins increasing the volume and frequency of his inhalations while shortening the exhalations. With the inspiratory reserve after a normal breath at or near its peak, and with the respirations becoming increasingly rapid, the diver is forcing his lungs to be full on every breath. Soon the exhalations become so short that the majority of his breathing is taking place at the top of the lungs, the inspiratory reserve. This entire problem is the result of improper breathing technique, which also causes a buildup of carbon dioxide in the system. Carbon dioxide is what stimulates us to breathe, so as its concentration increases, so does our breathing rate.

Maintaining full lungs while hyperbreathing causes a diver to be additionally buoyant. Divers can hold six to ten pints of air in their lungs while hyperbreathing, causing a six- to ten-pound increase in buoyancy, since one pint of air equals one pint of buoyancy. It therefore becomes extremely easy for a diver to rise a few feet, possibly promoting lung injury.

Long, slow exhalations, two to four times the length of the inhalations, are the key to breathing properly. Don't exhale more air than you inhale, just take more time to do it. This process will lower the amount of carbon dioxide in the body, and therefore reduce the respiratory rate and the urge to breathe, as well as lower the heart rate. Consequently, fatigue is also decreased. Most importantly, proper breathing will greatly decrease the chances of lung overexpansion injury.

5. While managing buoyancy. Any diver not skilled in breathing normally on scuba or performing safe buoyancy control techniques is at greater risk of making breath-hold ascents. Unfortunately, most divers don't truly understand good buoyancy control. They believe that it means you should fill your lungs to rise and empty them to sink. It's true that this will make you go up and down, but for good buoyancy control, you should always maintain normal breathing. With proper buoyancy, when a slight change in depth is desired, simply inhale or exhale a little more air, then wait a second or two (still breathing normally) for the buoyancy change to occur. For the most effective control, changes in body postures and movements are also necessary. With proper technique, only the subtlest of changes in exhaling and inhaling with be necessary.

6. When overweighted. Most public safety divers use too much weight. Overweighted divers add air to their BCD to achieve neutral or slightly negative buoyancy underwater. Because one pound of buoyancy equals one pint of air, a diver with six extra pounds on his belt will need six pints of air in his BCD to neutralize the overweighting. In accordance with Boyle's law, an additional six pints must be added for every atmosphere the diver descends. At ninety-nine feet, the diver will have a surface equivalent of twenty-four pints of air in his BCD just to counteract the six-pound overweighting. With this amount of air in the BCD, every bit of rise will result in a significant increase in BCD volume and buoyancy. This problem can be observed, in that most overweighted divers will be unable to stop at any given level without moving. If they stop moving, they will either rise or sink.

One real problem with the overweighted diver is the increased risk of making an uncontrolled rapid ascent. If the diver experiences a problem and becomes stressed or panicked, he will have less ability to find stasis, given the dramatic buoyancy changes caused by a full BCD rising and sinking through the water column. A small rise can turn into a full-fledged ascent, to the point that the diver cannot control it if he is busy with some other problem. This is especially dangerous for divers who use their dry suits to control their buoyancy, since the result can be a carotid sinus reflex and unconsciousness during ascent.

7. When a diver accidentally drops his weight belt. Anytime a diver's weight belt falls off by accident, there is an immediate possibility of an uncontrolled ascent and lung overexpansion injury. At that point, all he can do is flare out to slow his ascent. The problem will be even worse if the diver was overweighted and therefore had too much air in his BCD when the weight belt fell off. If a diver makes an uncontrolled ascent, he should be pulled from underwater operations for twenty-four hours and thoroughly checked and monitored for any symptoms of diving maladies by someone trained to perform dive-related field neurological evaluations.

8. While carrying a body to the surface. Adult drowning victims generally weigh from eight to sixteen pounds underwater. A diver not using a lift system will have to bring the body up by adding a small amount of air to his BCD and slowly kicking to the surface. All along the way, he should be prepared to vent the BCD in the event that he drops the body. If a lift bag is not available, a safer method for the diver is to tie a line or strap around the victim and let support personnel raise him to the surface.

9. While riding a lift bag. As a general rule, public safety divers should use a lift bag to bring up objects that weigh over ten pounds. In practice, too often the temptation is to use a lift bag as an elevator, possibly without paying much attention to the ascent rate. Divers should not ride a lift bag to the surface. If you do make an ascent this way, be sure to vent the bag to keep your ascent rate down to one foot every two seconds. If the rate becomes too fast, let go of the bag and get clear the area above and below it.

10. When being pulled off the bottom. A tender must only take in the slack on a diver's ascent and avoid pulling at a rate faster than one foot every two seconds.

11. During underwater tow-sled operations. Tow sleds pull divers through the water at high speeds, invoking the chance of rapid depth changes. To minimize the risk of lung overexpansion injury, a tow sled must be pulled by a competent boat operator.

12. When hanging on to lines in rough water. Divers often use ascent and descent lines, especially for safety stops. If the surface waves rise above two feet, however, divers should avoid holding fast to those lines. Instead, they should ride up and down with the waves, monitoring their depth gauges and allowing the line to pass freely through their fingers.

13. While breathing heavily in response to finding the target. It's a natural reaction to become excited when you found what you've been looking for, but a diver's immediate duty before giving that 6 + 6 + 6 signal is to keep his breathing under control and concentrate on long, slow exhalations. Do not ascend until your breathing is back to normal.

14. When there are predisposing physiological factors. Any of the standard contraindications to diving, such as congestion, smoker's lungs, pulmonary disease, or asthma, can cause lung overexpansion injuries, even if the diver follows all of the rules. The problem is that air can become trapped within affected portions of the lung and then expand during the ascent. A diver should avoid underwater operations for two weeks after the last signs and symptoms of pulmonary problems, and should afterward dive only with the approval of a physician.

Types of Lung Overexpansion Injury

Although the causal mechanism is the same in all cases, the resulting injuries vary greatly, ranging in severity from mild discomfort to death. Familiarity with the signs and symptoms is important in determining whether a diver may need further evaluation.

Arterial Gas Embolism (AGE): An embolus is any clot, bubble, or clump that is lodged in and blocking, partially or otherwise, a blood vessel. Lung overexpansion incurred in diving operations can result in the introduction of gas bubbles into the pulmonary arterial blood flow. Specifically, air is forced from the alveoli into the surrounding pulmonary capillaries. These emboli can obstruct arterial blood flow to the heart and brain (cerebral arterial gas embolism, or CAGE), as well as to other parts of the body. The condition can progress to unilateral or bilateral paralysis, as well as cardiac and respiratory arrest. The symptoms can also be as subtle as minor tingling or an inability to maintain mental focus.

Another form of AGE involves the heart. In fact, AGE fatalities are sometimes incorrectly diagnosed as myocardial infarction (heart attack) due to the presentation and standard autopsy procedures. Such a victim did, in fact, suffer a heart attack, but the instigating cause was an AGE.

The heart can be affected by an AGE directly or indirectly. Directly, bubbles can enter the coronary arteries and cause a blockage, resulting in a heart attack. Air entering the left atrium and ventricles can adversely affect cardiac function by reducing the heart's ability to pump. Indirectly, bubbles blocking blood flow to the vasomotor center in the lower brain stem can disrupt normal heart action.

Blood pressure and pulse quality and rate should be carefully monitored and recorded. If a blood pressure cuff isn't available, monitor the radial pulse of a patient, since a minimum systolic pressure of 80 mmHg is necessary for a palpable radial pulse. A systolic pressure of 80 mmHg is indicative of hypotension and shock.

The signs and symptoms of AGE typically occur within the first ten to fifteen minutes after surfacing, and they are often similar to those presented in cases of heart attack or stroke. They include the following.

- Out-of-character behavior.
- Confusion.
- Anxiety or agitation.
- Denial.
- Unconsciousness.
- Dizziness.
- Unilateral or bilateral motor problems.
- Unilateral or bilateral paralysis.
- Numbness or tingling.
- Weakness.
- Loss of sensation.
- Chest pain.
- Headache.
- Nausea.

- Convulsions.
- Speech or visual disturbance.
- Gasping or shortness of breath.
- Dilated pupils.
- Heart attack signs and symptoms.
- Shock.
- Near-drowning or drowning signs, such as pink or white foam in the airways.
- Depression.
- Memory or concentration problems.
- Liebermeister's sign, a pale mottling of the tongue.

Pneumothorax: The etymology of this word (*pneumo* meaning air; *thorax* meaning chest) essentially describes what occurs in this type of injury. Air perforates the alveoli, escapes from the lungs, and is trapped in the pleural cavity, between the lungs and the inner wall of the chest. As the diver ascends, this air expands, pushing against the lungs, possibly causing partial or total collapse (tension pneumothorax). Because of pressure in the chest cavity, heart problems may also occur.

The signs and symptoms of pneumothorax include the following.

- Sharp chest pain, especially on deep inhalation.
- Sharp back pain, especially on deep inhalation.
- Shallow, rapid breathing (hypoventilation).
- Shortness of breath.
- Swelling of neck veins.
- Irregular pulse.
- Anxiety.
- Cyanosis of the lips and fingernails.
- Coughing.
- Shock.
- Uneven chest rise.
- Diminished lung sounds under a stethoscope.
- Deviated trachea, especially when swallowing.

Mediastinal Emphysema: The name of this condition also describes what occurs (*media* meaning middle; *stinal* referring to the sternum; *emphysema* being a condition of air trapped in body tissue). Gas escapes from overexpanded alveoli and enters spaces within the lungs, where it can track along airways and blood vessels, entering the mediastinum area between the lungs and around the heart. The real danger of this condition is that air trapped around the heart can decrease cardiac output.

In severe cases, there is also a danger that trapped air will exert pressure on coronary arteries, possibly initiating a heart attack.

The signs and symptoms of mediastinal emphysema are as follows.

- Any signs and symptoms of myocardial infarction.
- Weakness.
- Faint heart sounds.
- Tachycardia, or a pulse of greater than 120 beats per minute.
- Hypotension, or low blood pressure.
- Crepitation, or a crackling sound accompanying the action of the heart, as heard through a stethoscope.
- Swelling in the neck.
- A full feeling in the throat.
- Retrosternal pain under the breastbone, possibly spreading to the neck and shoulders.
- Breathing difficulties.
- Cyanosis.
- Shock.

Subcutaneous Emphysema: In this type of injury, air escapes from the alveoli and follows a similar route as in cases of mediastinal emphysema. With subcutaneous emphysema, however, the wayward air enters the regions under the skin of the neck and upper chest wall above the clavicle, or collarbone. This is the least severe type of lung overexpansion injury, and it is typically left to resolve itself after a physician verifies that no other form of injury is present.

The signs and symptoms of this condition are as follows.

- Swelling under the skin in the neck and collarbone region.
- Breathing and swallowing difficulties.
- Voice changes, usually tinny-sounding.
- Crepitation, or a crackling sensation upon touch of the swollen areas, with a feeling similar to that of bubble packing.

Decompression Sickness

According to Henry's law, as the pressure on a specified gas in a mixture increases, the more that gas will enter solution. During normal respiration at the surface, nitrogen, which has no physiological function or use, simply passes in and out of the body as we breathe. When a diver breathes air, however, the increased partial pressure of nitrogen in his lungs consequently increases the pressure of nitrogen in the blood, where it dissolves into solution. Carried by the bloodstream, the dissolved nitrogen begins to diffuse into tissues throughout the body, including muscle, fat, the brain, the bones, and the organs. This journey of nitrogen from the diver's regulator into his tissues is called ongassing. The process is accelerated by depth and time spent at depth. The longer a diver stays at depth, the more time that a greater amount of nitrogen has to enter his tissues. The more a given tissue is perfused with blood, the greater the amount of nitrogen it will receive, and the faster it will approach a state of saturation. Vascularization, tissue type, and exercise all contribute to the amount of nitrogen a tissue can absorb.

As the ambient pressure decreases during the ascent, the nitrogen dissolved in the tissues becomes less soluble. Nitrogen under a high partial pressure will naturally seek its outlet through an area of lower pressure—a situation analogous to carbon dioxide dissolving out of solution after the cap has been removed from a bottle of soda. In a diver, during an ascent, this pressure differential is first realized within the lungs. As blood with a high partial pressure of nitrogen passes over the alveoli, now containing a lower partial pressure, nitrogen will diffuse outward and be expelled from the body through exhalation. Because the process is slow and steady, the blood will retain some nitrogen in solution, and some will even be reabsorbed at the ensuing inhalation. If the pressure is released from a diver's body too quickly, however, as during a rapid ascent, the transition of nitrogen from a dissolved to a gaseous state will not occur strictly within the lungs. Rather, nitrogen bubbles will form within the tissues and blood of the diver. These bubbles can create a host of different problems by blocking blood vessels and cutting off circulation, putting pressure against the nerves in the body, and physically damaging delicate tissues, including the brain and spinal cord. The symptoms of DCS can appear underwater during the ascent, immediately after the dive, or many hours later. Typically, the first symptoms present within the first two hours after surfacing. The results can range from simple skin irritation to numbness, tingling, joint pain, paralysis, and death.

Although most public safety divers don't follow dive profiles that would normally lead to a high buildup of nitrogen, decompression sickness is a possibility after any excursion underwater. Thus, divers should be monitored for any postoperation problems they may be experiencing, especially if any safety practices were violated.

Types of DCS

Because it can affect nearly every system of the body, decompression sickness can take on a variety of forms. Type I DCS refers to conditions that involve pain only, skin bends (pruritis or other cutaneous circulation abnormalities), the lymphatic system, or fatigue. Type II DCS is far more serious, since it presents with neurological or cardiorespiratory symptoms.

It isn't clearly understood how gas bubbles damage the brain and spinal cord. There are probably both direct mechanical injuries and indirect damage from the immune and clotting responses of the body. Damage to the frontal lobe of the brain can cause changes in personality and behavior, while injury to the prefrontal area can cause emotional disturbances and complex motor disabilities. Damage to the cervical spine can result in full paralysis from the neck down, respiratory problems, or subtler symptoms, such as tingling, weakness, and numbness. Local sensory and motor capabilities can be affected if injury is sustained along any of the peripheral pathways. Early, subtle symptoms in the fingers, for example, could indicate cervical spinal cord damage, presaging severe symptoms from the neck down.

The general signs and symptoms of DCS include headache, fatigue, chest pain, dizziness, breathing difficulty, blurred vision, unconsciousness, convulsions, abdominal pain, bilateral paralysis, loss of bladder control, out-of-character behavior, joint pain, numbness, and tingling. Note that the joint pain, numbness, and tingling won't usually feel better or worse when touched or rubbed, as would an injury caused by trauma. Note also that the spectrum of symptoms listed above is extremely variable, depending on the location of the nitrogen bubbles in the body.

Because minor symptoms are often forerunners of major ones, you should never ignore tingling, weakness, or numbness. Take a few minutes to figure out what else could be the cause. Remove all circulatory obstructions, such as straps, exposure suits, and harnesses. If massaging, drying, rewarming, and moving the affected area don't fully alleviate the symptoms in less than five minutes, then the investigation and treatment must focus on more insidious causes.

Anything that increases the rate of ongassing and decreases the rate of offgassing can be considered a risk factor for DCS. Some of the primary risk factors are as follows.

1. *Ascending too rapidly.* The rapid, uncontrolled relaxation of pressure on a diver's body is the root cause of decompression sickness.

2. *Diving too deep for too long.* All divers are taught to use some type of dive table during their first certification course. Ideally, tenders should also have a basic knowledge of how to use these tables. Such a skill can be useful when reporting information to the Divers Alert Network (DAN) or a receiving hyperbaric facility. It can also prove helpful when making decisions regarding patient management. Still, presenting signs and symptoms are more important than a dive profile, since conservative profiles can lead to DCS, while a risky profile may result in no physiological problem whatsoever.

Although public safety divers rarely use computers, ask to see it if one is available, since it will recall, at minimum, the diver's last profile. Ask whether the computer was turned off or lost power before any of the dives. Ask whether the diver used the computer on every dive. Ask whether the diver was wearing it himself or whether a buddy wore it. Finally, check the computer for rapid ascent alarms.

If no computer was used, check the dive profiles, including those of repetitive dives. The profiler should have recorded the times and depths, and he should have been able to assign a dive table letter group for the diver. For repetitive dives, such information should be kept in one place; i.e., by the dive coordinator. If no computer was used and no profiles can be provided because the operation was run poorly, then assume that the tables were violated.

3. Too many repetitive dives and diving too many days in a row. Those who make three or more descents a day for a week or more, with at least some of those descents to below sixty feet, could be at increased risk. Public safety divers don't usually operate for that long a period; however, during a large-scale operation, as in the case of an airliner recovery, members could easily be pressed into serving well beyond their normal limits. Divers in such a scenario might well be asked to make over twenty-one descents in a week's time. It's recommended that each take one full, twenty-four-hour day off midweek to allow the body to offgas, hydrate, rewarm, and rest.

4. Dehydration. This is probably the second greatest risk factor for decompression sickness, ranking only behind violation of dive tables. If a diver is more hydrated at the beginning of a dive than at the end, then there will be more blood volume to facilitate ongassing than offgassing. Dehydration means that there is less blood to transport the nitrogen in solution from the tissues to the heart and lungs, where it can be returned to gaseous form and exhaled from the body. Dehydration also means that there is a smaller volume of body fluid to hold the nitrogen in solution. If the volume of the solvent decreases while the amount of solute remains constant, then the solution will automatically move closer and closer to the point at which nitrogen bubbles will form. A diver not only undergoes a reduction of body fluids during a dive, he also simultaneously increases the amount of nitrogen in his body.

5. Cold stress. The results of cold stress go hand in hand with dehydration, since it results in diuresis. Coldness also decreases peripheral circulation. Thus, during that portion of the dive when a diver is active and not very cold, the peripheral tissues are well perfused with nitrogenated blood. At the end of the dive, particularly during the ascent and hang time, a diver is more likely to suffer from cold stress and peripheral vasoconstriction, making offgassing more difficult.

Diving in cold water demands special precautions for the prevention of DCS. In addition to staying hydrated and wearing proper exposure protection, they must be more conservative with diving profiles. When using tables, divers in water of 70°F or less should treat their dive as though the bottom time were one letter group more than it actually is. Those divers who use computers must be more conservative by avoiding the computer's suggested no-decompression by an even greater margin than normal.

6. Exertion underwater and high air consumption. It is believed that exertion may actually promote the formation of tiny nitrogen bubbles. Although these bubbles may not themselves cause any DCS problems, they may instigate the formation of those that do. For example, bubbles form in the lubricating fluid of the joints as a result of normal joint action.

Exertion increases the perfusion of blood to working tissues, such as muscles. Perfusion is believed to be positively correlated with ongassing and offgassing.

Exertion can also result in a greater consumption of air, meaning increased fluid loss and heat loss with each breath. The denser the air, the greater the effect, so overexertion at greater depths can increase the magnitude of the dehydration and cold stress. Also, exertion typically occurs during the main portion of the dive, at the deepest depths, and during ongassing. Then, during the ascent, when the diver should be offgassing, typically he's making minimal movements and is breathing at a decreased rate and volume.

7. Altitude changes before or after diving. Altitude is a consideration for divers who have recently or will soon make a descent. The current minimum standards for flying after diving require that a diver wait at least twelve hours after making a single no-decompression descent, and longer than twelve hours if he made a decompression descent or multiple descents within the preceding twenty-four-hour period. However, medical research has found that cases of decompression sickness have occurred when flying as much as seventeen hours after diving, and DAN recommends a waiting period of at least twenty-four hours after any dive. To be safe, wait a minimum of twenty-four hours after diving before flying in a pressurized cabin, and wait longer if the flight is to be made in an unpressurized aircraft. Flying not only places a diver in an environment of lowered ambient pressures, it also involves several dehydrating processes. If the dive is to be made after a flight, wait at least six hours to allow the body to rehydrate and fully rest from traveling. In the mountains, divers should wait at least twenty-four hours before moving to higher elevations. Consider a 1,000-foot elevation to be the threshold at which high-altitude procedures become mandatory.

8. Postdive skin diving. Free diving usually involves repeated rapid descents, ascents, and breath holds. The periods of breath holds impair offgassing, and the rapid ascents could cause nitrogen to come out of solution if the diver has a high level of residual nitrogen. Also, simply being in the water can further decrease hydration and body temperature.

9. Alcohol. As we all know, alcohol impairs judgment and increases risk-taking behavior, neither of which are desirable for any aspect of public safety operations.

Because it is a diuretic, alcohol also causes fluid loss and dehydration, which in itself is a DCS factor.

Beyond these problems, alcohol reduces the surface tension of water. Theoretically, it might increase the size of asymptomatic DCS bubbles, making them symptomatic, injury-causing bubbles. Finally, alcohol can also mask signs and symptoms of the condition, which means that a diver with DCS might attempt another dive instead of seeking medical attention. When interviewing a potential DCS patient, ask, “How many alcoholic drinks have you consumed in the past twenty-four hours?” not, “Have you consumed any alcoholic beverages?” The thrust of the former may increase the chances of receiving a more truthful answer.

10. Taking a hot shower or bath after a dive. Very warm water can cause sweating and loss of body fluid, increasing the likelihood of DCS. More importantly, it may also increase the number of tiny, asymptomatic bubbles in the bloodstream, which could promote the formation of larger ones.

11. Fatigue. Lack of sleep was reported in one-third of all decompression illness patients in the 1992 DAN *Report on Decompression Illness and Diving Fatalities.* It is possible that the correlation between fatigue and DCS isn't due to the fatigue itself, but associated factors, such as alcohol, caffeine, and stress.

12. Obesity and poor health. Because adipose tissue (fat) has a high capacity for absorbing nitrogen, it is believed that obese people may be more susceptible to DCS. People who are in poor cardiovascular health may also be a greater risk, since their limited cardiovascular system may impair tissue offgassing.

13. Patent foramen ovale. In the fetus, all oxygenated blood is received from the mother, via the placenta, rather than from the fetal lungs. Thus, the fetal heart has a hole, the foramen ovale, that allows blood to flow directly from the left to the right atrium, bypassing the lungs. Usually this hole will close shortly after birth, but in some cases it does not, in which case it is termed a patent foramen ovale. Since this breach of the heart wall may leak venous blood laden with decompression bubbles directly into the arterial system, bypassing the lungs, it is hypothesized that it could increase the risk or severity of decompression sickness.

Does this possibility mean that every diver should undergo a cardiac workup to check for this condition? No. More research is still necessary. Members who experience undeserved DCS or AGE may want to have a workup to decide whether they should continue diving. More importantly, such hidden, random factors as patent foramen ovale demonstrate why the chance of DCS should not be ignored simply because the dive profiles were well within the no-decompression limits.

Prehospital Care for Diving Maladies

Don't assume that responding EMS personnel, whether BLS or ALS, have any degree of knowledge in recognizing or treating decompression sickness. It is the dive team's responsibility to ensure that local EMS personnel are properly educated on the subject. The same holds true for the local hospital's emergency department. Check to see whether the department has the relevant treatment protocols, as well as whether its staff members are aware of the incumbent need for dialogue with the Divers Alert Network when confronted with a decompression-related malady. If not, a diver may needlessly wait in an emergency room before ever receiving hyperbaric treatment, losing precious minutes, hours, or days at the risk of permanent injury or even death.

Prehospital care is about making decisions, not diagnoses. The risk factors described in this chapter are only intended to focus attention on a subject. Use risk factors as clues toward starting an evaluation before the signs and symptoms become serious enough to be noticeable. If the risk factor was serious in a given circumstance, or if more than one was present, perform a field neurological evaluation and monitor the diver for at least one hour. Since diving profiles aren't reliable indicators of diving-related injury, decompression sickness can strike even when a diver is well within the limits of tables, computers, and acceptable ascent rates. Thus, you should also perform a field neurological evaluation if any potential sign or symptom of DCI is present.

Field Neurological Evaluation

A field neurological evaluation (FNE) is used to discover subtle signs and symptoms of diving-related injuries, not to make a diagnosis. It is different from a standard secondary survey, taught in first aid and EMS programs, since its aim is in finding seemingly minor or elusive neurological and behavioral abnormalities, whereas a secondary survey is typically concerned with trauma and relatively obvious problems. For example, a standard secondary survey involves a check of grip strength and perhaps the sensation of a finger or two. An FNE, however, explores the sensation of at least three fingers on both hands to check the functioning of cervical nerves 6, 7, and 8. It also involves a check of grip strength, the ability of the fingers to remain spread against inward resistance, the ability of the fingers to remain straight forward against downward pressure, and other such tests. An FNE is conducted repetitively to record the progression of a problem, whereas a complete secondary survey is typically performed only once.

If an FNE shows signs and symptoms that may be related to the central nervous system, consider requesting ALS transport, since such conditions can worsen and cause problems that are immediately life-threatening. A Type II DCS condition in the upper cervical spine could cause respiratory arrest just as cervical injury in an auto accident could.

Because the decision-making process is so important, at least some of the dive team members and EMS personnel responding to an underwater incident should have FNE training and certification. The rule of thumb to follow when evaluating any diving-related problem is a simple one: When in doubt, treat it as a dive injury.

First Aid Treatment

The hospital treatment needed for most decompression illness patients is recompression and oxygen administration in a hyperbaric chamber. The increased ambient pressure within the chamber has a twofold benefit. First, it forces more oxygen into solution in the diver's body, allowing better oxygenation to tissues that may have experienced reduced perfusion. Second, the increased pressure decreases bubble size and forces gaseous nitrogen back into solution, thereby reducing damage and blockages caused by the bubbles. Not many hospitals have hyperbaric facilities, however, and the nearest facility may be in use when needed, meaning that a diver must go elsewhere. Potential delays in treatment mean that prehospital care is all the more important.

The prehospital care for diving-related injuries is straightforward. As with any injury or illness, the first items to take care of are basic life support and treatment for shock as necessary. Beyond that care, the most important treatment for any diving-related condition is the administration of oxygen.

The steps for prehospital care for any type of diving injury are as follows:

1. Administer basic life support as required.
2. Have EMS and the receiving hospital call the Divers Alert Network emergency hotline.
3. Immediately deliver 100 percent oxygen by demand valve to a breathing patient, or by bag-valve mask or positive-pressure device to one who isn't breathing or who is breathing inadequately.
4. Keep the patient dry and warm. Remove his wet suit. Cold stress can exacerbate a decompression illness problem by decreasing blood volume and increasing a patient's oxygen consumption as he shivers.
5. Establish a calm and reassuring dialogue with a conscious patient.
6. Gather information, and keep the diver's records and profile with the diver.
7. Monitor and treat for shock.
8. Perform a field neurological evaluation every fifteen minutes, and record the vital signs.
9. In general, assume that any diving injury is a worst-case scenario of arterial gas embolism, requiring hyperbaric treatment as soon as possible. Transport the diver to the appropriate facility without delay.

The most important treatment for any diving-related condition is the administration of oxygen.

Although many EMS protocols require a nonrebreather mask for breathing patients, decompression illness patients are most helped by the highest percentage of oxygen possible. The difference in delivery between a nonrebreather mask and a demand-valve mask can be almost 25 percent, delivering 75 percent O_2 and 99 percent O_2 respectively. Thus, your protocols should state that, for any diving-related accident, breathing patients should be placed on demand-valve oxygen masks.

If the diver has been in cold water, have him breathe prewarmed, hydrated oxygen. This will reduce both heat loss and dehydration.

If the patient is breathing so weakly that the inhalation isn't strong enough to crack the demand valve open, then administer oxygen through a nonrebreather mask set to a flow rate of 15 liters per minute. If the patient is breathing too rapidly or shallowly, try to talk him into breathing more efficiently, asking him to concentrate on long, slow exhalations.

Don't confuse a demand valve with a positive-pressure button, which is used to mechanically ventilate a nonbreathing patient. Positive-pressure buttons are found on demand valves, but not all demand valves have positive-pressure buttons. If your local protocols have taken demand valves from your ambulances because positive-pressure valves are no longer used by EMTs, try to get one back on board for breathing victims of dive mishaps.

Although the nearest hospital may not have hyperbaric facilities, begin transportation to that hospital, since it does at least have the staff and equipment to handle more immediate life-support concerns. During transportation, advise the emergency room personnel that the patient has suffered an injury related to scuba diving. Simply describing it as a "diving" injury may lead to false expectations and treatment biases on the part of receiving personnel.

Take special care when the evacuation route involves altitude changes greater than 500 to 750 feet, whether by surface transportation or by helicopter, since these ascents and descents may compound the diver's problems.

If a pneumothorax patient finds it more difficult to breathe when lying down, then transport the patient lying on his side, with the injured side down, or in a position of comfort. Otherwise, elevate the patient to a lateral recumbent position. If a tension pneumothorax is present, paramedics in some counties can begin treatment by venting the pleural cavity with cannulation by inserting a large-bore needle or flutter-valve chest tube. If possible, use an ALS ambulance for transport.

If a patient with suspected decompression sickness has no nausea, consider giving him fluids to enhance his blood volume. This in turn will serve the offgassing of nitrogen and a reduction of symptoms. BLS personnel can administer fluids orally, providing a liter of fluid each hour for the first two hours of treatment. ALS personnel can provide fluids intravenously, using either Ringer's lactate 5 percent dextrose or normal saline. Do not, however, use plain 5 percent dextrose, since the sugar will be metabolized, leaving water that can worsen neural tissue edema. Note that a urinary catheter must be in place for patients who are unable to urinate.

The Trendelenberg position, in which a patient is laid on his left side with his hips higher than his heart and the heart higher than his head, was formerly recommended for decompression illness patients. In theory, that position might cause air to be trapped in the left ventricle to keep it from traveling to the brain. Practical application of the Trendelenberg position has been accompanied by several problems, however. For one, patients have sustained spinal injury when objects were shoved underneath them to elevate the back. Respiratory stress has been induced by radical elevation angles, when the patient was kept in the position too long, and when his respirations weren't monitored before or during the positioning. Bear in mind that this position puts pressure on the diaphragm as well as on the cardiopulmonary system. Radical elevations and excessively long durations have also brought on cerebral edema. Finally, in the case of patent foramen ovale, bubbles could be shunted through the heart and into the arterial system.

Because of these concerns, the Divers Alert Network and the major dive certifying agencies don't recommend the Trendelenberg position for their suggested protocols.

Rather, they recommend either a basic shock position, a position of comfort for those patients with respiratory problems, or lying a nauseous patient on either side to prevent the aspiration of vomitus.

Carotid Sinus Reflex

In the body, blood pressure is monitored by the carotid sinus receptors, which are located in the carotid arteries that run through the neck, supplying blood to the brain. Normally, as blood pressure increases, these receptors send a message to the brain's cardioinhibitory center, which in turn causes the heart rate to decrease. That decrease lowers blood pressure.

The carotid sinus receptors can be stimulated by external, physical pressure. Thus, if a diver is wearing a tightly fitting hood, or if the neck seal on his dry suit isn't properly trimmed, the carotid sinus receptors will be stimulated into slowing the heart rate, even if the blood pressure isn't actually high. This will result in an inadequate blood supply to the brain, which will cause the diver to feel light-headed and eventually lead to unconsciousness.

Carotid sinus reflex is the reason neck seals should be trimmed properly. If several divers on a team share one dry suit, its seals should be trimmed to fit the largest diver. Divers with skinnier necks should use separate neoprene neck collars to prevent flooding. This points out another reason why divers shouldn't use their dry suit as a BCD.

Ear Barotrauma

If a diver is unable to equalize but continues to descend, the tympanic membrane of the ear may be pushed in so much by the increasing water pressure on the outside and the decreased gas volume in the middle ear that it ruptures. Eardrum rupture will be preceded by sharp pain, followed by relief when the membrane gives way. Usually a diver will experience vertigo as cold water rushes into his middle ear. If you experience vertigo, stop and get your breathing under control. If visibility allows, focus on a gauge, your hand, or some other object. If you must vomit, hold your regulator in your mouth and vomit through it. Although the vertigo will usually abate as the water in your ear warms up, the dive should be aborted anyway. Although a ruptured eardrum will heal, the process should still be monitored by a physician.

Eardrum rupture is the most common type of ear barotrauma, but it isn't the only type. The round window is a flexible membrane on the cochlea of the inner ear. It can be ruptured when increasing ambient pressure and the venous back pressure of a forceful Valsalva maneuver combine to cause overpressurization. When the round window ruptures, a diver will experience tinnitus and vertigo that will not go away. Round-window rupture is a serious injury, and it must be treated by an ear, nose, and throat specialist.

To prevent round-window rupture, divers should make slow descents, beginning to equalize before they leave the surface and every foot or two thereafter. A diver who experiences any type of ear injury should be given a field neurological evaluation for possible diving-related maladies caused by a rapid ascent or breath holding, and he should be taken to a physician. If vomiting occurred, it's possible that he aspirated water or vomitus, so be prepared for respiratory distress en route to the hospital.

Sinus Barotrauma

Blockage in the sinuses can prevent them from equalizing during normal breathing. As the afflicted diver descends, he will typically feel pain in the forehead, between the eyes, over the teeth, or in the cheekbones. The air spaces of the sinuses are squeezed, and the lower-than-ambient pressure within them pulls fluid and blood out of the surrounding tissues. The fluid and blood, later forced from the sinuses by reexpanding air, can usually be seen in the diver's mask when he surfaces.

Although the sinus squeeze is extremely uncomfortable, it doesn't usually require medical treatment. However, it can cause delays in a diving operation. No diver with a cold, allergy, or stuffiness should ever be allowed to dive.

Reverse Blocks

Sometimes a diver with a cold or allergy will take some type of decongestant before making a descent so as to avoid any problems with ear or sinus barotrauma. Although this may allow the diver to descend, the decongestants can wear off during the dive, allowing stuffiness and blockages to return. When the diver makes his ascent, air becomes trapped in his middle ear or sinuses, expanding and with no place to go, creating the possibility of an outwardly ruptured eardrum or sinus damage. If possible, avoid taking decongestants when diving. Reverse blocks can also occur after strenuous attempts to equalize during a descent.

Study Questions

1. What causes lung overexpansion injury?

2. What is Boyle's law?

3. Why aren't skin divers prone to lung overexpansion injury?

4. Why is a scuba diver different from a skin diver with respect to the potential for lung overexpansion injury?

5. Name several situations in which a diver might unconsciously hold his breath or otherwise be prone to lung overexpansion injury.

6. What is an arterial gas embolism?

7. Often misdiagnosed, the signs and symptoms of arterial gas embolism are often similar to those presented by what other ailments?

8. The type of lung overexpansion injury in which air perforates the alveoli and is trapped in the pleural cavity is known as ________.

9. The least severe type of lung overexpansion injury is ________.

10. What is ongassing?

11. What causes decompression sickness?

12. True or false: Type I DCS refers to conditions that involve pain only, skin bends, the lymphatic system, or fatigue.

13. Type II DCS presents with ________ or ________.

14. True or false: The joint pain, numbness, and tingling caused by DCS will usually lessen when given immediate massage therapy.

15. Name several of the primary risk factors for DCS.

16. Prehospital care is about making ________, not ________.

17. An examination method used to discover seemingly minor or elusive neurological and behavioral abnormalities, as opposed to signs of trauma and relatively obvious problems, is called a ________.

18. What are the twofold benefits of treating decompression illness in a hyperbaric chamber?

19. What is the most important treatment for any diving-related condition?

20. True or false: If a patient with suspected decompression sickness has no nausea, you should consider giving him fluids to enhance his blood volume.

21. True or false: The Trendelenberg position, in which a patient is laid on his left side with his hips higher than his heart and the heart higher than the head, is no longer recommended for DCS patients.

22. What is the main reason neck seals should be trimmed properly?

23. What is the most common type of ear barotrauma?

24. If fluid and blood are evident in a diver's mask when he surfaces, it could be a sign that he has suffered ________.

CHAPTER 16

Contaminated Water

A simple list of some of the diseases that can be contracted from diving in contaminated water should be enough to convince any diver to follow decontamination procedures. Such diseases include pfiesteria contamination, hepatitis, typhoid, paratyphoid, cholera, and schistosomiasis, in addition to more mundane problems, such as diarrhea, vomiting, headaches, and blurred vision. Skin rashes, skin ulcers, and burns are common in water contaminated by chemicals. As if such problems weren't enough, public safety divers put themselves at the greatest risk by exploring where the contamination is worst: in the sediment along the bottom.

DETERMINING CONTAMINATION

Of the two basic types of contaminants, biological and chemical, the former are easier to detect. Biological hazards are found by culturing water samples for several days. This can be done in labs located all over the country and for relatively little cost. Detecting chemical contamination is more difficult if you don't know what chemical you're looking for. The equipment necessary for analysis, such as gas chromatographs and spectrophotometers, is far more expensive, and fewer facilities have it. Also, it is sometimes more difficult to filter enough water to get a detectable concentration of the contaminants.

Of course, dive teams should never ignore obvious clues of contamination. Diving near sewage overflows is accompanied by biological hazards. If divers find sealed drums on the bottom, it's time to get out of the water. On a site, it is up to the diving coordinator and the safety officer to determine the level of protection necessary. An incident commander should request a properly trained and equipped haz mat team whenever there is an indication that such materials may be present. During the preincident planning phase, officers should seek proper evaluations of the waters in their area. If certain waterways are simply too dangerous for underwater operations, then the team's SOGs should state that personnel will not dive there.

In point of fact, tides, winds, and illegal dumping mean that divers can never be sure that a given site is free of contamination. Always assume the worst, and dive as safely as possible. A common practice to prevent fogging in a mask is to spit into it and rinse it out in the water, but you should avoid this practice, since the water may be polluted. Microscopic droplets of contaminated water can enter the eyes and nose and cause problems. Instead, bring drinking water to use for defogging masks. Similarly, you should never clear condensation inside a mask by flooding it while diving. Avoid, too, rinsing your mouth in water while diving. Instead, drink plenty of water before making the descent so that dry mouth won't be a problem. Never swallow water to help equalize pressure in the ears, and never remove the regulator from your mouth except when switching to an alternate source of air. Divers have a tendency to remove the regulator while at the surface, which makes the ingestion of water inevitable and increases the chances of drowning. The mask and regulator, or the full-face mask, must not be removed until the diver has been properly decontaminated. Any regulators that breathe wet must be removed from service and repaired, since these would only promote contamination.

General Decontamination Procedures

There are six recognized techniques for decontaminating equipment and personnel.

1. Mechanical: physically wiping the contaminant away.
2. Dilution: using water to reduce the strength of the contaminant.
3. Absorption: using other materials to soak up the contaminant.
4. Degradation: altering the contaminant by chemical means.
5. Isolation and disposal: packing the contaminant into containers.
6. Disinfection: washing with specified cleaners.

The dilution method can be set up and performed by most public safety dive teams, perhaps with the assistance of the local fire department, until a proper decontamination team is on the scene.

The equipment needed to perform a proper basic decontamination of this sort includes several small, rigid-sided children's pools, scrub brushes with handles at least two feet long, several five-gallon buckets, sponges, and either a hose with a spray nozzle or water fire extinguishers. Also needed are a few watertight containers large enough to hold all of a diver's equipment. Set up a decontamination area that can easily be cleaned afterward, and secure a water runoff area before the dive. Tarps made of plastic are preferable to those made of canvas or other absorbent fiber materials. If the diver used the same ground tarp to dress up and gear down, then that should also be decontaminated.

Because divers and tenders are susceptible to heat exhaustion during washdowns, which may keep them in their suits for twenty minutes or so, the washdown area might preferably be in the shade. Similarly, during cold-weather operations, use warm water if possible to reduce the risk of cold stress and hypothermia.

During decontamination procedures, all surface personnel should wear proper exposure suits so that they will be protected from contact with divers and wet gear, as well as splashing. Respirators, eye protection, and gloves are required.

On exiting the water, a diver must release his weight belts and fins, and leave them by the water's edge. His regulator, however, must remain in his mouth from the minute he enters the water until he has been cleared to remove them by the decontamination officer or his designee. If the water has degraded the silicone of the diaphragm enough that it has been compromised or no longer functions, the diver should carefully place a clean regulator second stage in his mouth for the remaining procedures. It is important that members practice nonverbal communication so that divers don't need to remove their regulators except for out-of-air emergencies.

The first step in the process is for the diver to step into one of the small pools and be rinsed off with copious amounts of clean, fresh water to remove the contaminant. This washdown should take a minimum of four to five minutes. At the same time, the decontamination team can scrub down the diver and his gear using brushes and a 10-percent bleach solution diluted in water. For petroleum-based contaminants, use a standard degreasing solution of dishwashing detergent and water. The specific agent to be used should be specified in the team's SOGs. As an alternative to mixing solutions, your team may opt to use a sprayer bottle, such as the kind commonly used with a garden hose to spray backyard fertilizers and pesticides. Set the dial to the maximum concentration of solution.

Clean the diver from the top down, rather than from the bottom up, and pay particular attention to any irregularities that may trap contaminants, such as zippers, seams, gloves, sheathes, pockets, and the like. During the washdown, the spray nozzle should be held at a downward angle from one to three feet from the diver. This will serve to reduce splashing. At no time should the nozzle come in contact with the gear or diver, or it too must be considered contaminated.

The next step is for the diver to remove his BCD and tank so that these items and the back of his exposure suit may be cleaned as well. During this phase, the diver should retain his regulator and keep his mask in place.

After being scrubbed and rinsed, the diver can remove his mask and regulator, then exit the pool, leaving his tanks and BCD in it. He should also remove his gloves and booties, then step into the second pool for a quick rinse with another cleaning agent. If the diver is wearing inner gloves, there should be still another washdown in which the inner gloves are cleaned, then removed, and the diver's hands cleaned separately. If the diver will be needed to participate in further dives, he can remain in his suit after removing his tank during the first washdown. However, he should not be allowed into the designated clean area, but rather, he should remain in the warm zone of the incident.

If a diver will be needed to participate in further dives, he can remain in his suit after removing his tank during the first washdown. (*Photo courtesy of Frederick E.Curtis.*)

After that rinse, the diver should remove his suit. If he was wearing a wet suit, he should enter a final pool for a complete washdown. This pool should be protected from view, since the diver will be stripping to his skin. If he was wearing a dry suit, this step won't be necessary unless there was failure of the seals and the diver was exposed to contaminated water.

The diver should then be given a clean, dry towel and a clean change of clothing. The towel is to be used once and then thrown away. Dispose of it properly, in a bag labeled with the type of hazardous material involved. Check with a trained haz mat unit for proper disposal procedures.

After all decontamination procedures have been followed, the diver should have a postoperation medical evaluation. After that exam, he may be considered decontaminated, although he should still wash his ears and take a personal shower with soap and water. Taking a personal shower should be considered standard procedure for public safety divers, particularly those who use wet suits. Thorough washing can reduce infection from bacteria, yeast, and fungi.

Most dive teams require their tenders to wear gloves while handling the tether lines. However, these gloves are usually either leather work gloves or fire gloves, both of which tend to absorb water, thereby exposing tenders to the same contaminants as the divers. Tenders should wear two pairs of medical latex gloves under appropriate work gloves, as well as a pair over them. In most cases, a vinyl or latex oversuit is all that will be required to protect their bodies from splashing and wet, contaminated ropes. Eye protection and respirators are also necessary. If a tender has been exposed to contaminated water, he must go through the same washdown procedure as the in-water team.

Suit Decontamination

As mentioned in Chapter 6, wet suits are inadequate for diving in contaminated water. These suits allow the water to contact the diver's skin directly, and they can act as a sponge by absorbing contaminants. Neoprene can pose additional dangers during washdown, since it can be degraded by certain contaminants, such as hydrocarbons. It's possible that a high-pressure spray could force contaminants through the degraded neoprene, directly to the diver. In fact, it may take gasoline just over two hours to permeate neoprene.

Vulcanized rubber dry suits don't degrade easily, and they can be cleaned of most contaminants by the procedures described above. Moreover, they don't contain any cracks or crevasses that could trap pathogens and other contaminants. A vulcanized dry suit should have, at minimum, a latex dry hood, dry gloves, and heavy-duty dry boots. If the water is littered with debris that might puncture a suit, the diver should wear some type of protective coverall over the dry suit, and he should also wear puncture-resistant gloves over his dry gloves.

Ear Infection and Wounds

Ear infection is a common problem in diving, whether the waters are clean or polluted. Because the hood seals off the ears, the amount of bacteria in a diver's ear can rise dramatically within a very short while. The increased humidity and temperature of the ear continue to make these bacteria a problem even after the operation has concluded. Plugging the ears with cotton while diving in polluted water only adds to the problem, and it may hinder equalization. Earplugs are now available that are designed for diving. They allow a diver to equalize while keeping contaminated water out of the outer ear canal. These may be helpful for divers who wear wet hoods.

When wearing a dry suit, the hood should have good seals and a proper underlying headpiece that will allow for equalization and a tiny amount of ventilation for the ears. Using eardrops after the dive may help to reduce otitis externa, or outer-ear infection. Avoid rinsing the ears with salt water, since the salt crystals will keep moisture in them for extended periods. Instead, rinse them with clean, fresh water and then with a fifty-fifty mixture of alcohol and acetic acid (which will help to maintain a proper pH) for three to four seconds. Rinsing for a longer time can deplete the cerumen, or earwax. Over-the-counter solutions are available, but their effectiveness in conjunction with diving operations hasn't yet been tested. At least one physician recommends filling the ear with olive oil or baby oil before making the dive, since studies indicate that these benign coatings will make the ear more resistant to infection.

After rinsing, the outer ear should be towel-dried. If possible, gently dry the ear canal with a flow of heated air, as from a hair dryer. Never clean the ears with swabs or any other tool, since doing so will remove protective earwax and scratches inflicted in the skin can increase the chances of outer-ear infection. Consult an ear, nose, and throat doctor as soon as possible if you suspect a problem, and of course, never dive with an ear infection.

Diving in polluted water with wounds increases the chances of infection, and it should be avoided at all costs. If you must proceed with operations, cover the wound with a petroleum-based antibiotic ointment before the dive, and wash with soap and water afterward to remove the ointment. If you are wounded during a dive, clean it thoroughly with iodine solution, scrubbing if possible and irrigating thoroughly.

Decontaminating the Victim

Always bear in mind that the victim has had prolonged exposure to the same contaminants that threatened the diver, yet without any protective equipment.

One of the primary concerns for the care of a contaminated victim should be in the establishment of a secured airway, whether by endotracheal intubation by ALS crews or bag-valve mask by BLS crews. Either way, a quick suctioning of the oropharynx can remove most, if not all, of the contaminant from that area. If the victim is completely covered with contaminant, quickly wash him down using sponges from the decontamination kit and a detergent-and-water mix.

Local emergency protocols may dictate the proper prehospital methods to follow. The receiving facility must be notified in advance of the delivery of a contaminated patient. There are several methods of decontaminating victims in the field. The Fire Department of New York uses commercially available, one-time-use stretcher tubs that fit in the top of standard ambulance stretchers.

When handling a contaminated victim, EMS personnel and other patient handlers should follow the same safety measures as tenders, including coveralls, gloves, and face protection. Also, their rig will need to be fully decontaminated according to local protocols.

STUDY QUESTIONS

1. How are biological hazards detected?

2. Why is chemical contamination more difficult to detect than biological hazards?

3. What are the six basic ways to decontaminate equipment and personnel?

4. The decontamination method that can be set up and performed by most public safety dive teams is ________.

5. When should a diver take off his fins and weight belt?

6. When may a diver remove his mask and regulator?

7. True or false: To prevent contamination caused by handling the tether line, the authors recommend that tenders protect their hands by wearing two pairs of medical latex gloves under appropriate work gloves, as well as a pair over them.

8. True or false: It may take gasoline just over two hours to permeate neoprene.

9. Why should you not rinse your ears with a salt water solution after making a dive?

Chapter 17

Handling the Drowning Victim

Think about how Hollywood and television portray the drowning victim. Usually the victim is seen thrashing about at the surface, calling for help and splashing violently as he struggles for life. If no one hears him, or if the hero doesn't arrive in time, the victim suddenly stops moving and sinks under the surface. The last thing the audience sees is his hand groping skyward as it disappears from view.

In reality, drowning is a silent event. A drowning person no longer has the ability to scream, raise his arms or head out of the water, or struggle about. A victim might descend to the bottom within seconds without resurfacing, or he might bob for up to a minute or so, but in either case, there is little struggling and noise.

Adults take an average of sixty seconds to drown, and children take an average of twenty seconds. The majority of drownings are witnessed, and one or more witnesses are usually within 150 feet of the victim. Why don't more of those witnesses try to save the victim before it's too late? One of the principal reasons is that people don't realize what is happening. They don't know what a drowning looks like, having gained a false impression of it from television and the movies.

Dry Drowning

Both wet- and dry-drowning victims die of fatal cerebral hypoxia caused by asphyxiation. However, unlike a wet-drowning victim, who suffocates because his airway is loaded with the water he drowned in, the dry-drowning victim's airway is

Drowning is a silent event.

blocked either by respiratory reflex or by physical obstructions of the bronchi, such as mucus, foam, or vomit.

About 10 to 20 percent of all drownings are believed to be dry drownings. The typical dry drowning occurs when a person falls into the water, gasps, aspirates a small amount of water, and then experiences a laryngospasm, a reflexive response that seals off the airway. The shock to the body of sudden immersion in cold water is part of this response. Any aspirated fluid irritates the bronchial lining, which responds by secreting a thick, protective mucus. Hypercapnia, or an excessive amount of carbon dioxide in the blood, induces the body to breathe, but with the airway closed, blood fluid from the pulmonary capillaries, rather than air, is drawn into the damaged alveoli. This fluid destroys surfactant, causing collapse of the alveoli. The mix of aspirated water, mucus, and blood produces a white foam. If enough blood is in the mixture, the foam will take on a pinkish cast. The laryngospasm relaxes shortly before death; however, hardening of the mucus into plugs and a possible bronchiolar spasm continue to prevent water from entering the lungs. Thus, the drowning is dry.

The victim of a dry drowning is essentially in a state of suspended animation. Vasoconstriction in response to the cold sends a surfeit of blood to the core. The peripheral tissues experience a rapid drop in temperature, causing a drop in cellular metabolism, resulting in a lower need for oxygen. The core temperature eventually drops as well, protecting the vital organs from anoxic cellular death.

Dry-drowning victims have the greatest chance of being saved by aggressive rescue efforts. Resuscitation efforts are aided by several factors, including the freer gas exchange that is possible within relatively dry lungs, blood pooling in the core, and delayed cellular death from hypoxia.

Wet Drowning

In a wet drowning, the victim is usually already in the water but for some reason cannot swim to safety. Although he may stay above the surface for some time, he eventually weakens due to fatigue, cold, or a combination of the two. As he weakens, he sinks and reflexively holds his breath. The need for air will impel him to the surface just long enough to take a breath. He won't have the ability to hold his face out of the water long enough to both take a breath and scream. Thus, the drowning is silent.

The victim will sink again and rise again, but soon only his mouth and nose are the only parts to break the surface. At some point he'll no longer be able to bob to the surface and so will sink to the bottom. Eventually the urge to breathe will cause him to ingest large quantities of water into his stomach as he tries to hold his breath. He may vomit in response, and when breath holding is no longer possible, he'll gasp and aspirate water directly into his lungs. There will be a second phase of apnea, then a period of involuntary gasping underwater, sometimes referred to as agonal gasping, that can last for several minutes. Respiratory arrest occurs when the gasping ceases. Cardiac arrhythmias and cardiac arrest follow, and brain death is the last step in the process.

Because of the manner of drowning, as well as physical damage to the lungs, victims of wet drowning have far less chance of being resuscitated than victims of dry drowning. Still, since such a diagnosis cannot be made at the scene, any drowning victim should be given the best resuscitation efforts possible.

Near Drowning

Sudden facial contact with cold water (defined as being below 70°F) can trigger a response called the immersion reflex, formerly called the mammalian diving reflex. This series of bodily responses sharply reduces blood circulation to most parts of the body except the lungs, heart, and brain. Breathing stops and the heart rate is greatly reduced. That reflex will allow any oxygen remaining in the blood to be used by the brain, where it is needed for survival. Although the blood may have little oxygen, the brain may still survive. Because it has been cooled by immersion in the water, its need for oxygen has been reduced. The cooler the water, the greater the victim's chances for survival.

Suppose a victim or even a member of the dive team almost drowns and is pulled from the water coughing, choking, or barely breathing. After a few minutes of recovery, he seems fine and may even feel fine. Still, that person should not simply be allowed to go home. In a near drowning, a person does aspirate water. Although he may never lose consciousness or stop breathing, some of the same changes can occur in the lungs as occur in victims of wet drownings. These can lead to tremendous complications, possibly death. Furthermore, any aspirated water may contain bacteria and may therefore cause lung infection. Thus, if anyone has a near-drowning experience, he must be sent to the hospital to be checked out, monitored, and possibly given antibiotics as a preventive measure.

Victim Handling in the Water

Dive teams often become so focused on making an operation successful that they forget the second part of their job, which is to take care of a victim once he has been found. Too often, victims are recovered from the bottom only to be yanked around like luggage until they make it into the back of an ambulance. Only the most inept of public responders would pick up a cardiac patient by the armpits and drag him to the EMT vehicle, yet that is precisely the sort of behavior too often displayed when drowning victims are hauled up out of the water.

Mishandling has several root causes. Rescuers who aren't trained to work in the water may not be prepared to handle cold, wet, slippery victims. Rescuers often aren't physically fit for the task or may underestimate the amount of exertion required. They may be cold-stressed, hypothermic, and without proper hand protection. The lack of appropriate SOPs and the relative rarity of water-related emergencies are also significant contributors to the problem.

On discovery of a victim at the bottom, and after taking the requisite moment to calm down, the diver should pick up the victim, making every effort to keep the victim's body horizontal. Generally, the victim will be face-down, and the diver should keep him in that position until reaching the surface. This will prevent any more water from entering the victim's airway. In the rescue mode, the diver should also, if possible, seal the victim's mouth and nose during the ascent.

On surfacing, the diver should gently roll the victim face-up. Although many divers are trained to administer in-water breathing to a drowning victim, they should not do so in this case. In the early days of rescue/recovery dive teams, divers were taught to initiate mouth-to-mouth breathing on surfacing so as to induce EMS personnel to continue basic life support efforts and possibly revive the victim. In those days, EMS personnel often didn't initiate BLS on long-term drowning victims. Today, after so

On surfacing, the diver should gently roll the victim face-up.

many documented cases of successful saves, the use of an in-water ventilation protocol is no longer necessary or recommended. The victim will be a maximum of 150 feet from the tender, and the diver is on a tether, so transport to safer environs won't take long. Also, in-water rescue breathing may allow water to enter the victim's airway, especially as the diver is being pulled toward shore. The primary argument against the practice, however, centers around the diver's safety. Removing the mask and regulator to administer rescue breathing while in the water leaves the diver at risk of a mishap also. It also exposes the diver to possible contamination, and means that the diver must make mouth-to-mouth contact with the victim with no protective barrier in place.

As the diver is being pulled toward shore, he should continue to make every attempt to coddle the victim, covering the victim's nose and mouth to prevent water from entering the airway and lungs.

Passing the Victim to Shore

The way in which the diver hands the victim over to shore personnel depends on the type of shoreline or platform involved. It may even be necessary for the backup diver to deploy on the surface and with all equipment in place to assist the primary diver.

If the shore has a gradual slope or if the operation has been staged from a low platform, the diver should bring in the victim as close as possible. Shore personnel should then gently place the victim onto a floating backboard or transport device, or they should at least pick up and carry the patient to a transport device waiting at the water's edge.

Roll-Up Straps

If there is no gradual slope or low platform, a different method is required. Without the proper tools, it is extremely difficult to get any victim into a transport device or boat, much less one who might weigh in excess of two hundred pounds. As in a movable pulley rope system, roll-up straps allow you to gain about a 2:1 mechanical advantage over the weight of the victim, and they reduce the risk of injuring him as well.

To make roll-up straps, use a pair of two- to three-inch webbing straps approximately fifteen feet long, and put a No. 4 grommet in one end of each. Depending on need, a standard carabiner can be inserted into each grommet.

To use the straps, one end of each must be secured to the boat or platform. The carabiners can be hooked under the gunwale or secured to any other firm anchor

Roll-up straps being used during a training exercise.

point. If necessary, members can even stand on the ends of the straps. With the near ends anchored, pass the straps under and around the victim, then back up to the rescuers' hands. One of the straps should pass midway between the victim's elbows and shoulders, and the other should pass midway along his thigh. His arms should be lying flat alongside his body.

It is imperative that the straps be checked before hauling. The returning end of a strap must lie directly over the section leaving the boat. If the strap returns to the boat at an angle, it will attempt to straighten itself out as the victim is rolled into the vessel. The will cause the strap to travel along the victim's body. If this occurs, the strap could end up around the victim's neck, hanging him, or the victim could fall out of the arrangement entirely. Also, make sure that the webbing isn't twisted. If a rescuer is in the water, then he is the best person to check the placement of the straps.

Although one rescuer can pull up a full-sized adult when using the strap method, the hauling procedure is much easier when performed by two. When two rescuers are pulling, the person at the head is in control. The rescuers should stay low in the boat, and they may need to counterbalance a small vessel to keep it from capsizing. Once the victim is near the gunwale, the head rescuer should reach over and secure his head, protecting it from harm and helping to maintain an airway. The victim is then gently brought on board, possibly onto a transport device. In some cases, it may be desirable to lay a backboard directly on the gunwale or at the edge of the platform. If so, be sure to use a backboard with enough lift to float the victim in case the boat capsizes during transport.

Other tools besides webbing can be used for the roll-up procedure, such as cargo netting and plastic snow fencing. Whatever you use, practice first on land and then in the water.

No matter how the victim is raised, he must be kept horizontal. Because of the cold and other factors, the victim's blood will have pooled in his core. If you pick him up in a vertical position, the blood in his core will be pulled by gravity into his legs, thus depriving the core and brain, possibly causing shock or even death. Gently place the victim on a transport device and carefully move him to the ambulance. Since many drownings involve swimming and boating accidents, the backboard should be equipped with built-in straps for securing the patient, as well as head and neck stabilizers to help immobilize the spine. Even if no accident was involved, you should never carry a drowning victim by the arms and legs, or by any other similarly informal means.

Prehospital Care

Treat for hypothermia. In fact, many drowning victims are hypothermic even before they enter the water. Although a drowning victim may look completely pale and dead, it doesn't mean that there isn't any hope for him. Any pronouncement of death should be made by a qualified physician at the hospital, not anticipated by emergency responders at the water's edge.

Whether breathing or not, a drowning victim has already lost so much heat to the water that he simply can't afford to lose any more. Thus, if at all possible, drowning victims should not be placed directly on a cold backboard. No matter what material it's made of, it will conduct significant heat from the patient. Be sure to secure a wool, fleece, or other noncotton blanket on the board as insulation. In some districts, wool may not be allowed in proximity to oxygen administration or because of its allergy potential, but there are other options. Without insulation of some kind, the backboard can conduct more heat away from the patient than the air above.

If a revived patient is placed on an uninsulated backboard, he's likely to shiver, which will increase his oxygen consumption by 50 percent. If his ability to take in and transport oxygen is already decreased due to trauma, shock, or other factors, then shivering might be the straw that results in cardiac arrest. At the same time, an immobile patient will become colder much faster, since he can't use voluntary muscle movement to generate heat. To make matters worse, a patient experiencing hemorrhagic shock is also much more likely to become hypothermic.

When a patient is removed from the water, his skin and clothing are soaking wet, so heat loss can become even greater by exposure to the wind. Protect the drowning victim as soon as possible with wool blankets. Because they work by reflecting heat, space blankets should only be used as wind protection outside of wool blankets. If the patient is hypothermic, there is little or no heat to reflect. Worse, space blankets have no insulative or absorbent capabilities. One of the best tools for keeping a drowning victim warm is a heat-retaining recovery suit.

As soon as the patient has been placed on a backboard, immediately begin standard support procedures. Perform a primary survey and administer basic life support as indicated. Because the heartbeat of a hypothermic patient could be very slow or irregular, take the pulse for a full minute before performing CPR. Ventilate the patient with sufficient amounts of oxygen. Insufficient ventilations increase the chances of further alveoli collapse and pulmonary edema. Use prewarmed, hydrated oxygen if possible and if your local protocols allow.

Once the patient is in the ambulance and out of the elements, cut off his wet clothing and dry him gently to help him warm up. Be sure, however, that the temperature inside the ambulance isn't too high. A high ambient temperature will cause peripheral vasodilation, resulting in a loss of warm core blood to the periphery and a drop in blood pressure. This may set the stage for afterdrop shock, a life-threatening condition. Additionally, increasing the temperature of peripheral tissues will increase cellular metabolism. Since those tissues aren't receiving adequate supplies of oxygenated blood, however, cellular death will also increase.

Advanced Life Support

The vasoconstriction common to hypothermia patients may make it very difficult to start an intravenous line, and this can result in delays in the field. Even if a line can be started, most drugs require a certain body temperature in order to be effective. If the patient is severely hypothermic, the drugs will most likely not produce the desired effect. Paramedics must be particularly careful if they administer multiple doses, given that they might all kick in to produce an overdose effect once the patient warms sufficiently. Also, if rewarming does take place in the field, then the patient's blood volume will simultaneously be increased by the IV fluid and the fluid returning from the tissues, the interstitial space. The result could be cerebral edema and pulmonary edema, the latter of which in particular is a tremendous problem for any drowning patient.

Intubation is a violent event that could put a hypothermic patient with a weak heartbeat into ventricular fibrillation. When intubating a patient, be careful to stimulate the vagal nerve area of the neck as little as possible.

The effects of defibrillation on a very cold heart are still not fully understood. Follow your local protocols for defibrillating hypothermic patients.

Pulmonary edema, which is found nearly without exception in drowning victims, is a major contraindication for using military antishock trousers (MAST). Thus, MAST should never by used on drowning or near-drowning victims.

As mentioned above, a drowning accident may also involve head, spinal, or other injuries. Paramedics should be diligent in performing a thorough secondary survey.

The emergency department may require additional information from the dive team, including the following.

1. The victim's age.
2. The type of accident.
3. Other injuries in addition to drowning.
4. Prehospital care administered, including CPR and the time that it was commenced.
5. The length of time underwater.
6. The temperature of the water.
7. The type of water, whether fresh or salty; clean or contaminated.
8. The victim's level of consciousness on return to shore and the presence of a pulse or spontaneous breathing.

Sinking and Decomposition

Although an unconscious person will sink, he may not do so immediately. Laryngospasm may have trapped enough air in his lungs to keep him buoyant for a while. Only after the laryngospasm relaxes will the victim expel air and sink.

Because fat floats and muscle sinks, the higher a person's muscle-to-fat ratio, the heavier the body will be underwater. If a person is wearing shoes, a watch, a wallet, jewelry, or a belt, he'll be even less buoyant. Clothing can also greatly affect whether a person sinks. Down coats, hip waders, rain gear, and windproof jackets can all trap air, delaying submergence.

Note that infants have a very low muscle-to-fat ratio, and therefore a low wet weight. This low wet weight, combined with plastic diapers and baby wraps that could trap air, can allow a baby to remain floating on the surface for quite some time. If an infant is suspected to be a drowning victim, make sure that surface personnel thoroughly check the shoreline and the water's surface.

Once a person dies, his body immediately begins to decompose. Underwater, a body decays through several internal and external processes. A basic understanding of these processes can help you determine what will happen to the body next, including whether or not it will float.

Putrefaction

Putrefaction is the decay of animal or plant tissue, especially proteins. It produces foul-smelling compounds, such as methane, ammonia, cadaverine, and putrescine. Decay involves the ingestion of tissue by bacteria that come from both the body itself and the water in which it is immersed. The bacteria ingest soft tissue, turning it into liquid and gas. The body then leaks the putrefied juices and tissues from the various orifices. The upper layer of skin blisters and peels off the body. Gas gives the body a swollen appearance, as well as buoyancy.

Putrefaction usually begins in the digestive system, where coliform and clostridial bacteria are plentiful. With gas production, the abdominal cavity begins to swell. Decomposition gases can cause the eyes and tongue to protrude out of their orifices, giving the corpse an appearance of having been strangled. Abdominal organs may be pushed out of the vaginal or rectal orifices.

As putrefaction continues, hemoglobin is decomposed into greenish compounds that stain the skin. Veins close to the skin will appear greenish in color, producing a skin pattern known as marbling. The remaining skin will turn to purple, to brown, and eventually to black.

Anthropophagy

The term anthropophagy literally means "human eating." It occurs when a human body is eaten by scavengers such as shrimp, crabs, fish, turtles, or other animals. It can result in a particularly gruesome sight that can disturb even the most seasoned recovery diver. Anthropophagy can also severely impede autopsy fact-finding. Typically, the scavengers begin eating at skin that is already broken, such as at the wound site, and they can quickly destroy evidence of the wound itself. Otherwise, the scavengers will seek conventional entrances into the body, such as the mouth, nose, and rectum. The damage they inflict may prevent the body from ever floating to the surface. Anthropophagy can begin minutes after a body reaches the bottom, although clothes are typically a barrier to scavengers for the first twenty-four hours. Still, immediate recovery is critical for a corpses in water that contains scavengers.

Skin Maceration

In water, skin will become macerated, making it soft, white, and wrinkled after a few hours in cold water, less in warm water. This condition is especially seen where the layers of keratin are thick, as in the palms and soles. Keratin is a protein found in human skin, hair, nails, and tooth enamel.

After a few days underwater, these layers of skin will begin to separate, a condition called washerwoman's fingers. After one to two weeks, depending on the temperature of the water, the skin will then separate from the hands and feet. It is important to recover this hand skin, if possible, in the event that fingerprinting later becomes necessary.

Reflotation

The time it takes for a body to produce enough gas to rise off the bottom depends on many variables, including the temperature of the water. The warmer the temperature, the greater the putrefaction and production of gas. Bodies in 75°F water can show advanced stages of putrefaction in forty-eight hours. Bodies in cold water, just above freezing, may experience minimal putrefaction and never produce enough gas to rise. Cold layers within the water column may hinder a rising body also, since the cold may cause the gases to recondense, stealing away buoyancy. If a body is in deep water, the effects of pressure may be so great that a sufficient amount of gas may never come out of solution. The deeper a body is, the less likely it is that it will ever produce enough gas to float, simply because the gases it produces will dissolve back into the surrounding tissues and the water itself. Thus, the bodies of many who die in deep waters are never found.

The salinity of the water is another factor, since salt water is denser and provides more buoyancy. Contamination is also important. The more bacteria in the water, the more there are to go to work on the victim. A body by a sewage outfall should decay and float quickly, whereas a body in a pristine lake should take much longer to rise. Stagnant water will usually produce faster rates of putrefaction than clean, moving water.

The last meal that the victim ate can also play a role, since some foods tend to produce more gas than others. The bacteria that give us gas problems are the same ones that produce gas in corpses. If scavengers besiege the body, however, they may consume all of the soft tissues, leaving little left to decay. Also, they may consume enough of the corpse that there is little left to trap any gases.

Finally, newborn infants don't have the same internal bacteria or gastric juices as older individuals, so the rate of putrefaction in an infant's body is minimal. For this reason, infants may not float to the surface if they ever manage to sink in the first place. The number of variables involved means that tables and charts showing the flotation times and rates at various water temperatures are highly inaccurate.

When the body gains a little buoyancy, it will usually rise with the spine up, its limbs hanging down, although obese people and some women may float face-up because of abdominal and breast swelling. If there is a current, the buoyancy from decay will cause the body to travel along the bottom until it gains enough buoyancy to rise completely off the bottom. As the body moves, it will incur abrasions as it hits rocks and other debris during its journey downstream. Travel abrasions will typically be found on the forehead, nose, elbows, knuckles, knees, tops of the feet, and the toes. Fractures may also occur if the body is traveling in swift water of about six knots or more. Snagging on debris in swift water can cause a limb bent by rigor mortis to straighten out. Also, if a body is being carried by fast water, any clothing will likely be ripped off. Don't assume that the recovery of a nude corpse means that the person went in without clothing. Search around for any traces of garments, and try to determine the path that the body might have taken, including where the victim first entered the water.

Dive teams often state that ninety percent of all drowning victims come up in the same general area. After the body surfaces, it's usually found in an area frequented by people, such as a park, a good fishing spot, or a swimming hole. The principal reason that the bodies are found in such places, however, is because that's where people congregate. In reality, the body may have drifted along the surface for quite some distance before reaching a point where it could most readily be seen.

As a body's digestive tract fills with gas, it ascends to shallower water. During the ascent, the ambient pressure exerted by the water decreases. Consequently, the gases within expand all the more. As the gases expand, the body expands, displacing more water and becoming even more buoyant. At some point in shallow water, the gases expand so much that the rate of ascent causes the body to pop to the surface, much like a diver who ascends by adding air to his BCD.

This first ascent to the surface is called primary flotation, and it is mainly caused by gas produced in the digestive system. When the body reaches the surface, it will most likely be positioned face-down. At this point, some gas may escape from the various orifices, causing the body to lose its buoyancy and sink back down.

When it sinks back down to the bottom, it will continue to decompose, possibly never again to resurface. Alternatively, it may rise again after the extremities have putrefied enough to become swollen. This second resurfacing, characterized by grotesquely discolored and swollen limbs, is called secondary flotation. After secondary flotation, the body will typically remain on the surface until it decomposes much further. A body in secondary flotation may float face up or face down. It will be very buoyant, and the limbs will be buoyantly floating as well. The skin will be very dark purple or black due to the latter stage of putrefaction and perhaps exposure to the sun.

Bagging Corpses

When a team is operating in a recovery mode, any bodies found should always be bagged before being brought to the surface. Bagging is important for several reasons. First, it preserves the integrity of the body, which may be falling apart if it has been

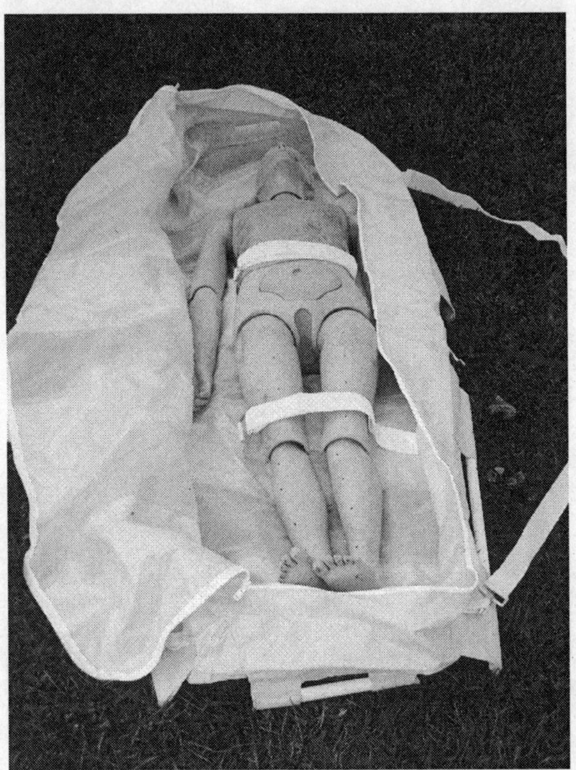

A body bag should always be used to recover a body at depth.

underwater for some time. It can also help retain important forensic evidence if the person was missing or involved in a crime. Bagging a body to conceal it from public display can also be psychologically important for response personnel, and it is particularly important if journalists and family members are on the scene.

Any diver who ventures underwater to recover a corpse may encounter a sight too grisly for description. That is a fact of our business, and one for which all recovery personnel must be prepared. The putrefaction of a human body is one of the many reasons why dry suits and full-face masks are preferred, and why wet suits should be destroyed after so much exposure to contamination. Gloves in particular should be destroyed after recovering a corpse. Because of the potential appearance of a decomposed body, a diver needs to remember that, if he must vomit underwater, he must hold on to his regulator or full-face mask and vomit through it.

Study Questions

1. Why is drowning a silent event?

2. On average, how long do adults take to drown?

3. Approximately what percentage of drownings are believed to be dry drownings?

4. What is a laryngospasm?

5. Why do dry-drowning victims have the greatest chance of being saved by aggressive rescue efforts?

6. What is agonal gasping?

7. In what position should a diver bring a drowning victim to the surface?

8. True or false: On reaching the surface, the diver should roll the victim face up and immediately initiate mouth-to-mouth breathing.

9. When using roll-up straps, where should the straps pass along the victim's body?

10. What is the basic prehospital treatment protocol for drowning victims?

11. Why might an unconscious victim not sink immediately?

12. Where does the process of putrefaction usually begin?

13. Define anthropophagy.

14. True or false: Because the solubility of residual gas decreases with temperature, bodies tend to float sooner in deep, cold, still bodies of water.

15. When a human body gains a little buoyancy, how will it usually rise?

16. What typically causes secondary flotation?

Chapter 18

Submerged Vehicles

Unlike an accident on the road, a mishap involving a submerged vehicle offers few or none of the usual outward signs that might serve an incident commander in making his size-up. Often there aren't any bubbles or even pools of fuel at the surface to indicate the exact location of the vehicle. Responders won't know the condition of the victims or how many are trapped. Lacking such operant clues, responding personnel often lose sight of what they need to do and how to do it. Too often, teams and personnel simply don't know what happens when a vehicle enters the water, and so remain largely unaware of the dangers posed by such operations.

The Star Project

In the fall of 1992, the Michigan State Police conducted a study, the Submerged Transportation Accident Research (STAR) project, in which twenty-five standard production automobiles were launched into the water to study their flotation characteristics. The study covered a wide range of vehicle types, from compacts to vans. None of the vehicles were altered in any way that would affect their behavior in the water.

Each of the vehicles was rigged to drive itself off a ramp and into the water. The time it took the vehicle to sink to the bottom, damage to the vehicle, and the viability of the electrical system were all quantified.

Although some of the vehicles sank faster and some slower, the average longevity at the surface was between two and three minutes. Buoyancy was affected by the condition of the windows and the integrity of the door and window seals. The windows of some of the vehicles broke on impact with the water. Another factor affecting flotation was the weight of the vehicle.

After so much time at the surface, the vehicles tended to sink nosefirst as the weight of the engine pulled them down. Most of the water entered through the air vents as air escaped through the backseats into the trunk or tailgate, and on through the trunk or tailgate seals. In many cases, the roof indented slightly from the lower-than-ambient pressure within the passenger compartment. Contrary to popular myth, the vehicles did not retain a breathable air pocket in the backseat or trunk areas.

As the vehicles sank, they descended vertically. In water more than fourteen feet deep, sinking vehicles landed on the nose and then tended to turn turtle, falling over onto the roof. Vehicles in shallower water tended to settle on their wheels. The electrical systems, including lights, windshield wipers, and windows, were operational for an average of ten minutes after submergence.

Perhaps the most important aspect of the STAR project was that it showed the potential for escape from a vehicle in the water. Because of the difference in pressure between the water and the air within a sinking vehicle, the doors cannot be opened. However, the rather lengthy flotation time should allow occupants to remove their seat belts, roll down the windows, and then escape by climbing through them. Even power windows should function for the first few minutes of immersion. Unfortunately, many people don't use that escape route. Instead, some believe that, in their panic, passengers will roll up the windows to keep the water out rather than using them as a pathway to safety.

The STAR project was followed by the STAR II project, again conducted by the Michigan State Police. STAR II was essentially the same type of study, except that it involved school buses. Sixteen buses of various sizes, ranging from twelve-passenger to sixty-five-passenger models, were driven off a ramp into the water. Entering the water at very slow speeds, less than eight miles an hour, the buses tended to float, lingering at the surface for as long as three and a half minutes. Other results were grimmer. For higher-speed entries, buses sank in less than twenty seconds. Smaller buses tended to sink in less time, as little as nine seconds. The problem was that the front windshield, mounted in a rubber frame and designed to be kicked out if necessary, immediately imploded on impact with the water. In all likelihood, the driver would be killed by that implosion. The accordion door of larger buses also collapsed. Without the windshield and door to hold the water back, the buses would quickly flood and sink. Smaller, van-type school buses had similar problems. Although these were equipped with standard hinged doors that remained intact, these buses experienced additional flooding through the engine cowling next to the driver, and therefore also sank quickly. As with passenger cars, in water deeper than the length of the vehicle, buses tended to land upside down. Little chance for escape was found for any model of bus.

The information obtained from these studies is useful to the public safety diver. Knowing the depth of the water and the type of vehicle involved will give a good indication as to whether the target is wheels up or not. This can be important in black water, saving a diver time in orienting himself to his environment. Also, if the electrical system is still operating, it'll indicate that the submergence occurred within the preceding ten minutes or so, raising the chances of making successful rescues.

Hazards of Submerged Vehicles

When a diver pokes his head, arms, or even his whole body into a submerged car, he enters one of the most dangerous environments in underwater operations. Even the roomiest vehicle will seem unduly cramped to a diver wearing full gear, and the roof and small portals limit both access to the surface and access by a responding backup. Steering wheels, shifters, hand brakes, seat belts, loose carpeting, and a host of other items present entanglement dangers, and the body of the car itself may have been deformed by its excursion off the road.

Moving around a submerged vehicle offers too much potential for entanglement of the tether line on the wheels, undercarriage, bumpers, exhaust pipes, and all else. The simple rule for traversing vehicles underwater is to go over, never go around. If the tether snags on an object, line-pull signals become ineffective, backup divers lose their route of direct access, and the primary can only make a direct ascent by severing the line.

Broken glass and twisted metal are hazards to be respected, and even shatterproof glass can cause lacerations. Simply reaching into a vehicle requires slow, cautious movements. Be wary of seat springs, which can easily puncture a dry suit, glove, or hand. Jagged metal is the greatest concern. Many divers have sliced open exposure suits and flesh alike while working on submerged vehicles.

Although such ordinary items as soda bottles and spare tires are normally harmless, underwater they can be dangerous, even deadly. A soda bottle dislodged from under a seat during a search operation can knock out a mask or regulator as it rockets toward the surface. A spare tire that pops out when a diver opens the trunk or rear door can have enough force to break the diver's neck. When searching any part of a submerged vehicle, stay low. If you must pop the trunk, get below it before you open the lid. Use extreme caution when searching inside the trunk as well. Even if the spare tire doesn't come out right away, it may be unsecured and simply waiting for you to dislodge it. Windshield-washer bottles and other items under the hood can also become underwater missiles.

Never get on the downstream side of an unstable vehicle or one resting in water flowing faster than about a quarter of a knot. The current will eddy on that side, and the eddying will dig a hole in soft bottom just past the car. The hole may eventually become deep enough for the car to fall into it, and any diver working there may be trapped.

One of the most dangerous yet least obvious hazards of submerged vehicles is contamination from petroleum products. Even in a short period, gasoline, oil, and diesel fuel can break down the neoprene and latex of many exposure suits. On contact with the skin, they can cause burning and irritation, and fuel in the eyes can cause temporary blindness. High-octane aviation fuel poses an even more extreme hazard. Ingesting petroleum-contaminated water can bring on lipoid pneumonia, a life-threatening condition.

Petroleum products will float to the surface. Divers can avoid the slick by ducking under the edge of it, but there is a better way. Certain liquids are available that act as repellents against some types of fuel. Check your local regulations to find out whether or not you may use such chemicals in your waterways, since they can be harmful to fish and other wildlife. At minimum, dishwashing liquid will also help disperse the slick, at least long enough to put a diver in the water. Dispense some liberally on the surface, and squirt down the diver as well. Having liquid soap on his dry suit will help dispel some of the petrochemicals he encounters, at least for a time.

A diver must be thoroughly decontaminated after any submerged vehicle operation. If an exposure suit is already rotting away from contact with petrochemicals, cut the suit right off of him and dispose of it as hazardous waste.

Gearing Up

Because of the abundance of chemicals, divers shouldn't acclimate their faces at the water's edge. Rather, they should use a bucket of cold, clean water.

Ideally, divers should wear full-face masks equipped with transfer blocks and vulcanized rubber dry suits. Besides wearing Kevlar® glove liners, they might also wrap duct tape around the fingers of the gloves for greater protection against abrasions and cuts. Lightweight plastic helmets are also recommended.

Hand Tools

Although it might come as a surprise, bungee cords are the most necessary and useful items to have when working on a vehicle underwater. Closed doors are extremely dangerous, and divers can keep them held open with some creative use of these rubber straps. Any door, lid, or tailgate should be secured as soon as you open it. Some divers place wedges in the hinges as well, in case the bungee cords are accidentally dislodged. If you use wedges in the hinges, first be sure that the door is open fully. Otherwise, the wedges could fall out if you subsequently open the door wider.

A window punch is a pen-sized tool that, when pushed against a glass window, will load and then suddenly release a spring. That spring imparts a jolt to a tapered metal point against the glass, shattering it. The same type of window punch used for land rescue can be used for submerged vehicle operations, but you may need to loosen the spring about a half-turn to improve the efficiency of the tool underwater.

A seat-belt cutter is a plastic hook with a razor blade inside the bend of the hook. When looped over a seat belt and pulled, it will slide through the belt. Because the blade is shielded, it presents minimal danger to the diver or victim. Shears may also be used to remove seat belts from trapped passengers.

On discovery of a vehicle, you may wish to float a tethered buoy from it to mark its location. However, if the line presents too much of an entanglement hazard, you may wish to forego the buoy in favor of a tether line to shore, or perhaps just a notation on the profile sheet.

Underwater communications systems are extremely helpful for working around submerged vehicles. Because such operations often require two divers to be under-

A diver preparing to descend on a vehicle. The line on the left is his hardwired communication link; the line on the right is a tether that he will attach to the vehicle.

water, being able to communicate will allow them to work in greater harmony. Also, if the divers are examining criminal evidence or using lift bags, they'll need to be able to communicate with their tenders more explicitly than basic line signals will allow. These systems also provide a greater margin of safety during those times when direct-line access is lost, as when a diver enters the vehicle.

Heavy-Rescue Tools

Being able to cut the roof off a car to extricate the victims more easily certainly sounds like a great idea. However, without excellent dive skills and extensive training, using a heavy-rescue tool underwater simply brings along too many hazards and problems to be worth the benefit. Even commercial divers, who sometimes use heavy tools underwater, must train with those tools for many months before using them on actual jobs.

If they aren't used with buoyancy collars, heavy tools will simply pull a diver down too quickly to be safe. For a diver to move the tool at the bottom, he must inflate his BCD, posing a risk of uncontrolled ascent if he and the tool part company. Heavy-rescue tools require good leverage, but in the weightless world at the bottom, a diver usually can't get the firm foothold he needs to operate such a tool effectively.

Perhaps the biggest problem with these tools underwater, however, is caused by restricted visibility. In black water, you can't even see where you place the tool or where the sharp, twisted metal is going once you set the jaws to work. You can't be sure that the victims aren't being harmed, nor can you be sure that your backup is out of the way.

Normally, in a rescue mode, trapped victims should be extricated by divers underwater, rather than removing the vehicle from the water and then pulling the victims out. However, there is one exception. If the passengers are pinned, it will be quicker, easier, and safer for all to lift the vehicle out and then cut them free.

Deploying to the Vehicle

Under certain circumstances, divers may not need to use tethers for submerged vehicle operations. After a descent line has been secured to the vehicle, if the divers are using an underwater communications system, if the visibility is over eight to ten feet, and if the sediment is of a type that won't obscure visibility if it gets stirred up, then the option exists to conduct the operation untethered. Multiple cutting tools and pony bottles remain mandatory, however.

Since entering a vehicle can snag or bend a tether, the backup diver should deploy once the target has been located. The backup descends by way of a contingency line and stands by underwater to assist the primary as needed. That way, there is a backup diver maintaining direct-line access at the opening of the overhead environment. At the surface, of course, the ninety-percent-ready diver should become a full backup, and another diver should suit up for the ninety-percent-ready slot.

Underwater Extrication

Extricating victims from a submerged vehicle can be extremely difficult. In some cases, the car doors will be locked or jammed shut. Use a window punch to break the side window, then try to unlock the door. If the door still won't open, you may try to extricate the victim through the window.

Before you attempt to pull any victim out of a vehicle, first be sure that you have cut off the seat belt. Cut it into several pieces to make sure that it'll stay out of your way.

If you aren't able to get the victim out through a window or door, lifting or towing the vehicle from the water may be the next option. Otherwise, you will have to try going in via the windshield or the back window. Because windshield glass is typically laminated, however, it will be difficult to remove underwater. Practice removing windshields on land before you attempt it below.

The victims may be anywhere inside the vehicle. In some cases, you may find the front seat empty. If the victims weren't wearing seat belts, they may have tumbled into the rear as the car rolled in the water. It's also possible that they panicked and followed the last traces of air into the backseat area. Always search the entire vehicle, including the rear deck and under the dashboard, until you are sure that there is no one else inside.

In the same vein, suppose you find a baby seat but there's no baby in it. It's possible that the child was left at home, but it's also possible that he was ejected from the seat. During your search, be sure to check all points above you, since disposable diapers can cause a baby to float.

Suicides and Homicides

For some reason, many people on the verge of suicide elect to take their vehicle with them. Washington, D.C., for example, has several in-water vehicular suicides each year. The tendency for these people is to lock the doors, roll up the windows, and race off an embankment.

Most vehicles involved in suicide attempts are driven into the water at high speed. Cars involved in homicides that have been staged to look like suicides usually wind up much closer to the shore, since they typically enter the water at slower speeds. Thus, a close-in vehicle with only minor damage may be cause for suspicion. Evidence of physical trauma found during an autopsy can often indicate whether an individual sustained injury as a result of vehicular causes per se or by some other external means. When a person goes through a windshield, there is little damage to the back of the head or the dorsal regions of the shoulders. When a person inside a vehicle drowns, there is little physical damage other than the standard impact marks sustained in the crash.

Always be aware that a vehicle in the water may involve a homicide, as in the infamous Susan Smith case, in which a young mother sent her car into a lake with her two young boys strapped in the backseat. That crime was not an isolated incident.

Many cars dumped in the water involve insurance fraud. When any suspicious circumstances are present, the vehicle should be thoroughly searched before and immediately after removal from the water. Pay special attention to the trunk, under the seats, and under the dash. All areas within twenty feet of the vehicle's location should also be searched. Bear in mind that as water rushes out of a vehicle, so will some of its contents. In cases of fraud, look especially for objects such as wooden blocks, rocks, or chunks of concrete that might have been used to depress the accelerator pedal. Inflatable toys may also be used for this; however, once underwater, such a toy will compress and probably float elsewhere. In at least one case, a dime was

Look for out-of-place objects. The brick on the floorboard, left, may have been used to depress the accelerator pedal.

inserted into the buckle of a seat belt, causing a woman to be trapped in a car driven into the water by her husband, who escaped unharmed.

After the vehicle has been towed out, divers can return to the bottom to search for more clues. In some cases, the entire route that the vehicle took may also need to be searched for evidence or additional victims.

Submerged Aircraft

Although it is rare that a dive team will respond to the scene of an aircraft crash, such a possibility always exists. If there is a chance of rescue, operations should proceed normally. If operating in a recovery mode, however, don't act too quickly.

Federal law CFR 14, NTSB 830 states that no one may recover an aircraft that hasn't been released by the Federal Aviation Administration or the National Transportation Safety Board. With that law in mind, it's best to play it safe and get permission from federal authorities before conducting any operations around a submerged aircraft.

If you must dive on this sort of wreckage, use the same procedures as you would for other vehicles. However, you must also exercise a greater degree of caution. Since

The wet-suit divers who worked around this crashed airplane suffered chemical burns from aviation fuel.

they're composed of lighter materials, typically aluminum, they'll usually be less intact than a car would be, and the wires and control cables inside the wings and fuselage may pose additional entanglement hazards. The risks associated with their fuels are many.

Public safety divers should not be asked to recover downed aircraft, and their SOGs should stipulate that they will not. The task of recovering vehicles from the water, especially aircraft, may require commercial divers. Airframes are more fragile than the chassis and bodies of land vehicles, and bringing them up from the bottom requires a knowledge of lifting points and techniques far beyond the scope of this book.

Recovering Vehicles

Public safety teams may sometimes need to lift heavy objects out of the water, as when recovering a vehicle to extricate the victims inside. However, it isn't the task of a public safety dive team to remove old cars, sunken boats, or other underwater trash. Removal of such items is a job for commercial divers or professional towing companies operating under OSHA standards. A dive team's SOGs should state that neither the team nor any divers on the team may be ordered to remove sunken junk. A team may legitimately circumvent OSHA, however, if it removes old vehicles as part of a bona fide training exercise.

Remembering Archimedes' principle, an object wholly or partially immersed in a liquid is buoyed upward by a force equal to the weight of the fluid displaced. Thus, the weight of a ship, or its tonnage, is expressed in terms of its displacement. One of the most common misconceptions about buoyancy is that objects underwater always weigh substantially less than they do on land. Objects of high-density material, such as steel, won't actually weigh much less in water, since the upward displacement force is small compared to the downward dry-weight force. In fact, the displacement force of dense objects is often so small that it is of little or no consequence when calculating the amount of lift needed to perform a recovery. For example, cold rolled steel weighs 490 pounds per cubic foot dry weight. Since one cubic foot of seawater weighs 64 pounds, a cubic foot of steel will weigh 426 pounds fully submerged (490 – 64 = 426). The seawater provides a lifting force of only 13 percent of the total weight of the object.

There are four basic means by which objects may be raised from the water. Dead lifting involves the use of mechanical lifting devices, such as cranes or tow trucks. Internal buoyancy lifts take advantage of voids and air spaces within the object, filling them with air to restore buoyancy. An object may also be made buoyant by pumping out the water that caused it to sink. Finally, attached buoyancy lifts involve using buoyancy devices affixed to the interior or exterior of the object.

Although simple in theory, dead lifting is a complicated process. Lifting by the wrong point can damage the object. Worse, the lift capacity of a mechanical device can

Although simple in theory, dead lifts should only be done under the supervision of a professional.

be severely affected by the current, high winds, cold temperatures, the length of the boom, the weight of the rigging, and the weight of the water trapped in the object. Underestimating any of these factors can have disastrous results. At minimum, dead lifts should only be performed under the supervision of a professional commercial diver, a professional rigger, or a professional salvager.

Internal buoyancy lifts should also only be done with professional supervision, since many hidden factors are involved in all phases of the lift. For example, the object may not be able to tolerate the outward pressures created by filling it with air. Also, to make a safe internal buoyancy lift, there must be some means of letting the water out. Vents that are too small may cause overpressurization, whereas vents that are too large will simply allow water to enter faster than it can be pumped out. Moving water, of course, has considerable force, and if it isn't kept under control, or if the buoyancy in a given compartment becomes too great, the object may roll, regurgitate air, and sink to the bottom again.

Pumping out is a method used to raise objects, such as ships, that are lying relatively close to the surface. To raise an object by pumping out the water in it, the cause of the sinking must first be repaired. This method rarely has applications for public safety operations.

Most divers are familiar with lifting by means of a buoyancy source, usually a lift bag, attached to the sunken object. Lift bags come in many sizes and shapes, with all types of features. Some divers attempt to make homemade lift devices out of steel drums, buckets, inner tubes, and other objects; however, using a homemade device isn't recommended, since it may fail or not work as predicted.

A number of problems accompany this method. To decide how many bags will be needed, you must first know the weight of the object. If no manufacturers information is available, you may be able to make a rough estimate of the weight based on the volume of the object and its material composition. Placing the lift bags incorrectly can result in more damage to the object than already exists. An object can be broken, stood on end, or capsized if the lifting force is in the wrong location. Bags must be attached to suitable lifting points, and those points must be able to support the maximum

Lift bags must be attached properly.

amount of force that the buoyancy device can generate. The choice of rigging materials is an equally complex matter for the same reason.

As the name suggests, an open-bottom bag is a teardrop-shaped item that is filled by placing an air line below it. If such a bag is disrupted, however, or if it lies over on reaching the surface, it will expel air and allow the object to sink again. Pillow and tube bags are shaped as their names suggest. They are fully enclosed, so the air inside cannot escape until it is released through a valve. Linkable bags are similar, but these can be connected together like sausage links. All such enclosed bags are equipped with overpressurization valves that work both manually and automatically.

Lift bags aren't rigid containers, so they are affected by Boyle's law, which states that pressure and volume are inversely related. As a bag rises, the air within it will expand, causing the object to rise faster. If the bags have been improperly attached or overfilled, or if any of a number of situations involving weight and center of gravity come into play, the outcome might well be the capsizing or resinking of the object. Such a worst-case scenario isn't meant to deter a team from learning how to lift vehicles by this method; it is merely another reminder that sufficient training is required to competently perform a deceptively simple task. As an introduction to some of the basic concerns inherent to these operations, a sample lifting problem appears in the Appendix.

The key to successfully raising an object by the attached-buoyancy method is in maintaining control.

As if the exigencies of lifting a heavy object out of water weren't enough, the mud on the bottom may make the operation even more complex. The force exerted by mud suction can exceed 100 pounds per square foot. Despite the temptation to do so, you should never deal with this resistance by adding more lift. When the suction finally breaks, it can do so quite violently, so that any additional force will contribute to a rapid and uncontrolled lift.

Mud suction can be overcome in a number of ways. One is to use a slow, steady pull, which may literally require hours before you see results. Basically, that slow pull will gradually allow water to fill the voids being created as the object is slowly lifted away from the bottom. Another method is to introduce air under the object. Similarly, you can also pump water underneath it. Water jetting can be used to loosen or remove portions of the sediment. Even rocking the object from side to side may help. The shape of the object has a large effect on the amount of suction present. The flatter the object, the greater the suction.

Study Questions

1. Based on the findings of the STAR project, approximately how long will the average automobile remain at the surface before sinking?

2. In water more than fourteen feet deep, how will an automobile tend to land, based on the findings of the STAR project?

3. What is probably the best method of escape from a sinking automobile?

4. Based on the findings of the STAR II project, what would most likely kill the driver of a bus that enters the water?

5. What may a diver infer if he discovers that the electrical system of a vehicle is still operating?

6. What general safety rule applies to traversing a submerged vehicle?

7. How should you open the trunk of a submerged vehicle?

8. Ideally, what underwater gear should divers wear for operations involving vehicles?

9. Name three reasons, as described in the text, why heavy-rescue tools might be unsuitable for underwater operations conducted by public safety teams.

10. Under what circumstances may divers operate around a submerged vehicle without a tether?

11. When should the backup diver deploy in submerged vehicle operations?

12. True or false: Cars involved in homicides that have been staged to look like suicides are usually farther away from shore, since the perpetrators typically drive them in at higher speeds.

13. What two federal agencies have jurisdiction over all civilian aircraft crash sites?

14. What are the four basic methods of lifting a submerged object?

Chapter 19

Boat Operations

Used as diving platforms, boundary markers, and work ferries, boats have an application in virtually any underwater operation. Using a boat should be mandatory when the dive zone is more than 125 feet from shore; when the water is moving faster than half a knot and there is no other stable platform for the tender upstream of the diver; or when the area to be searched has an overhead obstruction and direct-line access needs to be maintained. Boats should also be used in areas of heavy debris or vegetation, where a steeper angle of the tether line is desirable. If the shoreline is too steep, muddy, rocky, or impenetrable, stage the operation from a boat.

Choosing a Boat

In general, the best all-purpose craft for any dive team is an outboard-powered inflatable boat. The soft sides and low gunwales of inflatables are well-suited to divers, and they also make it easier to bring victims aboard. Soft sides are also less apt to be damaged by scrapes with docks, rocks, or other vessels. The light weight of an inflatable will allow personnel to carry it to and from the site, and if need be, the boat can be deflated for transport through the woods, then refilled from an air tank at the water's edge. Even though they're lighter, inflatables often have a higher weight capacity than rigid-hulled craft of comparable size. Because their means of flotation is along their outside edges, they are very stable in the water, and they tend to have good handling characteristics as well. An inflatable won't sink if you

In general, the best all-purpose craft for any dive team is an outboard-powered inflatable. (*Photo courtesy of Frederick E. Curtis.*)

capsize or swamp one, and it'll take you into areas where a vessel with a deeper draft can't go. A shallow draft is especially useful for shore launches. Since they're made of rubber, inflatables won't steal heat away from divers and victims as a metal craft will, and patching one is easier than making repairs to fiberglass or metal. An inflatable makes an excellent tending platform during operations on weak ice. Not the least of their merits is that they are less expensive than many other options of comparable function.

Despite all of these advantages, rigid-hulled boats are still quite useful, especially if you can keep your vessel docked in one location. Some teams opt for them if they work in swamps or other areas with many hidden obstructions. Also, occupants tend to stay drier in a rigid-hulled boat than they would in an inflatable, which can make a big difference in how divers and tenders feel. More stable at high speeds, rigid-hulled boats, especially fiberglass ones, can be extremely useful for teams that conduct operations offshore or on sizable inland waters. The likelihood of operating in contaminated water is an important consideration when choosing between basic hull designs.

A third option is to use an airboat, a flat-hulled vessel propelled by a giant fan in the rear. In the hands of a skilled operator, an airboat is excellent for use in very shallow water or even over ice. However, since they aren't able to carry much weight, they aren't the best diving platforms available.

Rigid-hulled boats can be advantageous for teams that conduct operations offshore or on sizable inland waters.

When purchasing a boat, teams must decide between jet drives and propellers. Jet drives offer safety advantages, since their moving parts are contained within the vessel. However, their intakes can suck debris into the impeller, which can be expensive to replace, and they have no true reverse—only a scoop that drops over the water jet, directing the flow toward the bow. Jet drives are also more expensive. Outboards work better for speed and deep water, although they must be used with even greater caution at a dive site. You may wish to put a prop guard over the propeller. A short-shaft motor and a drive linkage by way of a clutch rather than a shear pin are also probably preferable.

If there are only small bodies of water in your area, you may not even need a motor for the boat. Much will depend on its intended application and size. Many teams find that they can make a better diving platform by lashing together two smaller boats rather than using a single large one. If you have only one boat, it must be able to carry at least six people—the primary and his tender, the backup and his tender, the ninety-percent-ready diver, and the victim. Transportation is also an issue if you can't leave the boat moored in one body of water.

Before your team purchases any type of boat, check its rated maximum capacity on the small plate affixed to the transom. The indicated limit does not include the weight of the motor, tool kit, oars, safety equipment, anchors, and other standard items. Still, you must bear in mind that you can easily exceed the weight limit of a boat before you exceed the maximum number of personnel. A boat rated for four people or 700 pounds may not even hold three 250-pound firefighters. You must consider the entire load. All those tanks, regulators, BCDs, weight belts, and other items contribute significantly to the weight in a boat.

Equipping a Dive Boat

Any boat used by a dive team must meet or exceed Coast Guard regulations. Additionally, make sure that your boat has the following equipment on board.

- A PFD for everyone on board, including the victim.
- A Type IV throwable flotation device.
- Rescue rope throw bags.
- A first aid kit in a watertight container.
- For winter operations, wool or fleece blankets for the standby divers.
- Bottled drinking water.
- Fire extinguisher.
- A bullhorn or other signaling device, as well as a pea-less, nonmetallic whistle.
- Watertight light with spare batteries.
- Flares.
- Radio.
- Flashing lights, if permitted in your district.
- Diver-down flags.
- Roll-up straps.
- Pliers, vise grips, and adjustable wrenches, all with lanyards.
- One flat and one Phillips screwdriver, each with a lanyard.
- A folding pocketknife with a lanyard.
- Cable ties.
- Extra duct tape.
- A patch kit for an inflatable boat.
- Oars or paddles.
- A spark plug socket wrench with a lanyard, and spare spark plugs.
- A small can of spray ether for the carburetor.
- A starter cord for hand-cranking the motor.
- Three cotter pins and shear pins for the propeller.
- A spare propeller.
- Anchors and suitable line.

Anchoring

Searching an area on the bottom requires precise positioning and movement of the surface vessels, allowing divers to cover areas that barely overlap. One of the greatest problems that dive teams face is in keeping their boats in one place. If a boat or diving platform moves even a few feet, there will be a gap in the search pattern.

Even if the wind and current are minimal, a boat can still move. If a boat is moored by one anchor line, it'll be swung by the movement of the diver, especially if the tether line is longer than the anchor line. If the boat is swings back and forth, the anchor will pivot as well, and if the anchor loses its hold, the boat will drift.

The solution is to use a three-point hurricane anchoring system. By this method, the boat will have two anchors secured to the bow and one off to one side of the stern. Three anchors set this way will prevent the boat from moving, even with a current, wind, or a diver keeping a taut line in his search patterns.

To configure the anchors this way, maneuver the boat past the first point where you wish to begin searching, and deploy a buoy at that point. The buoy only marks

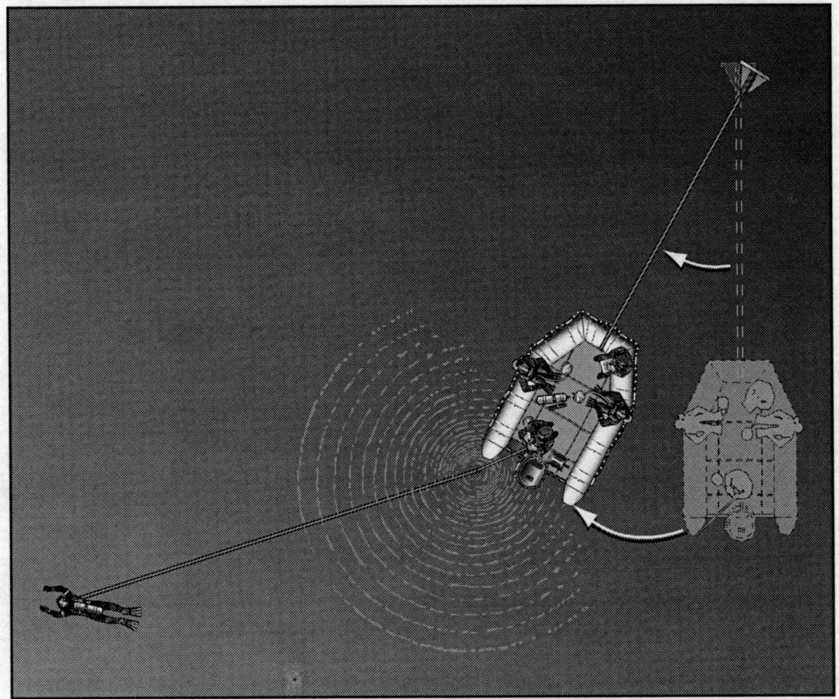

If a boat is moored by a single anchor line, its position will be influenced by the movement of the diver.

the position where the boat will be. It will be pulled in once the boat has been properly anchored. Next, position the boat so that the wind or the current, whichever is stronger, is at a 45-degree angle to the bow, and you are reapproaching the buoy from the downwind or downcurrent side. Before passing the buoy, drop the stern anchor, placing it on the same side of the boat as the buoy. Also, remember never to throw an anchor. To avoid tangling the line, anchors should always be lowered.

Continue moving past the buoy, paying out stern anchor line as you go. At some point past the buoy, drop a bow anchor on the same side as the stern anchor, so that the bow anchor is directly into the wind or current. Then, move the boat laterally, away from the stern anchor, to deploy the second bow anchor. Taking in and paying out line as necessary, set the boat at the focus of these anchors. Ideally, there should be an anchor 45 to 90 degrees off both sides of the bow and one off the stern. Don't shut off the engine until the boat is anchored securely.

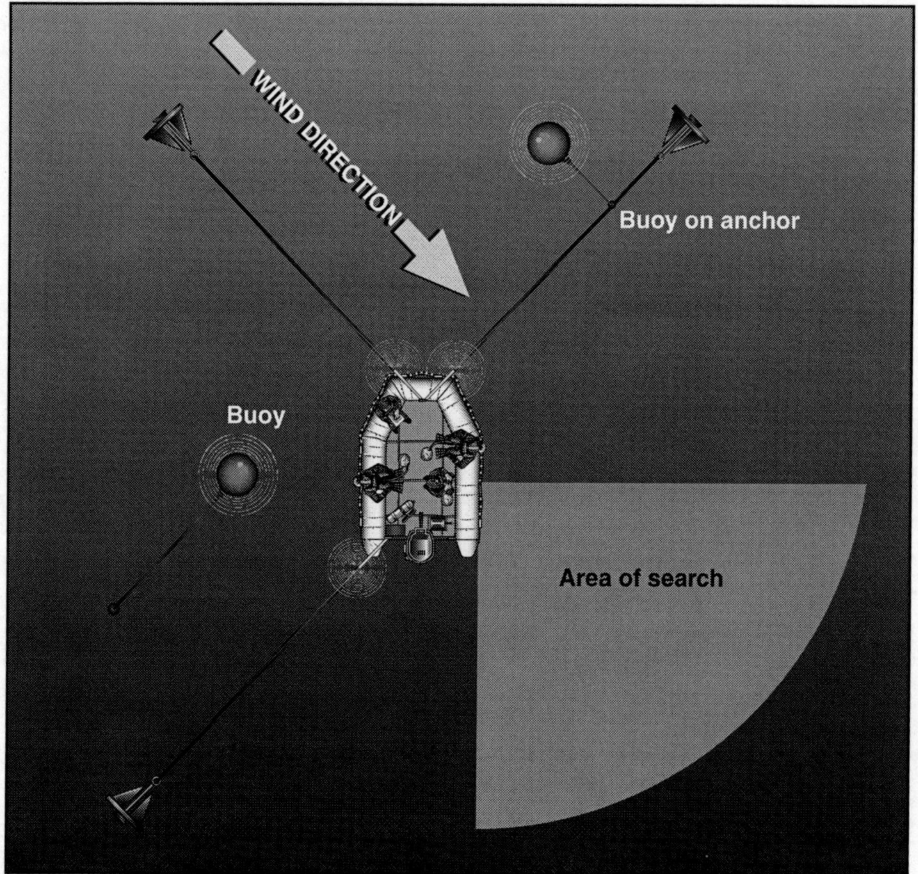

Three-point hurricane anchoring system.

To ensure that an anchor will hold, the anchor line paid out, or the scope, should ideally be seven times the depth of the water. This 7:1 ratio of scope to depth can be applied to actual operations as follows. Suppose that the depth of the water is 30 feet and that the diver is going to work off the starboard side of the boat. Drop the stern anchor 210 feet behind and well to the left of the buoy. Next, move the boat upstream or upwind approximately 420 feet and in line with the stern anchor to drop the bow anchor, also to the left of the buoy. Let the wind or current bring the boat back to a point between these two buoys, or use the engine, if necessary. Someone should take up line at the stern to prevent it from becoming entangled in the prop. Next, maneuver the boat to starboard and forward to drop the second bow anchor so that it is in line with, and approximately 210 feet from, the port bow anchor. The distance between the bow anchors will thus be equal to their scope. Finally, settle back into position so that the buoy is alongside the boat.

When you have searched all of the area possible from this location, change position by hauling in and giving out line as necessary. When you have exhausted the search range of these buoys, have another boat deploy a fourth anchor for you and bring you the end of the line. This anchor will replace one of the original three. Reset the remaining two anchors and continue the process of taking in and giving out line to move around the new, adjoining search area.

Although the three-point anchoring system is conceptually easy, it does require a great deal of practice to master. With practice, the entire process of hurricane anchoring can be accomplished in less than ten minutes.

Large-Area Searches

Once you are proficient in the procedures described above, you can integrate that system into large-area search patterns. Laying out a large area begins by dividing it into blocks. Using blocks about the size of a football field, a rectangle 300 hundred feet long and 150 feet wide, is one easily comprehensible way of visualizing the search zones.

Each block of an area to be searched can be marked with buoys. On the long side of the block, place four buoys 100 feet apart, numbered 1 through 4. Along the shorter side, where the team will start, place two buoys 75 feet apart. Designate these as A and B. For a longer or wider block, add additional buoys as appropriate.

To conduct the search, start the boat at the buoyed end of the grid, along the centerline. If enough personnel are available, divers can work simultaneously on both sides of the boat. Deploy them to just beyond the edges of the block, ensuring overlap with the adjoining block, and have them work inward, toward the boat. Once the divers have covered these sections, move the boat 25 to 35 feet farther down the centerline of the block. Repeat this procedure until the boat is even with Buoy 2. Then, deploy a diver to search the area that has been underneath the boat during the pre-

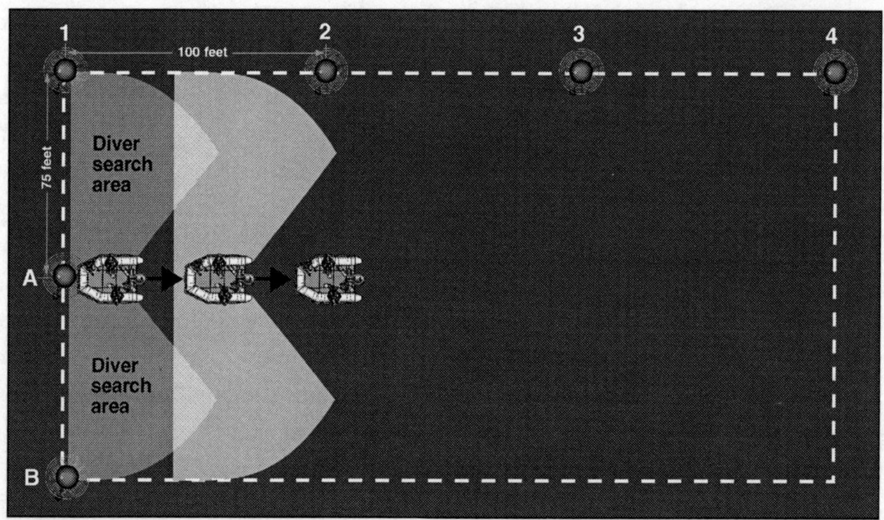

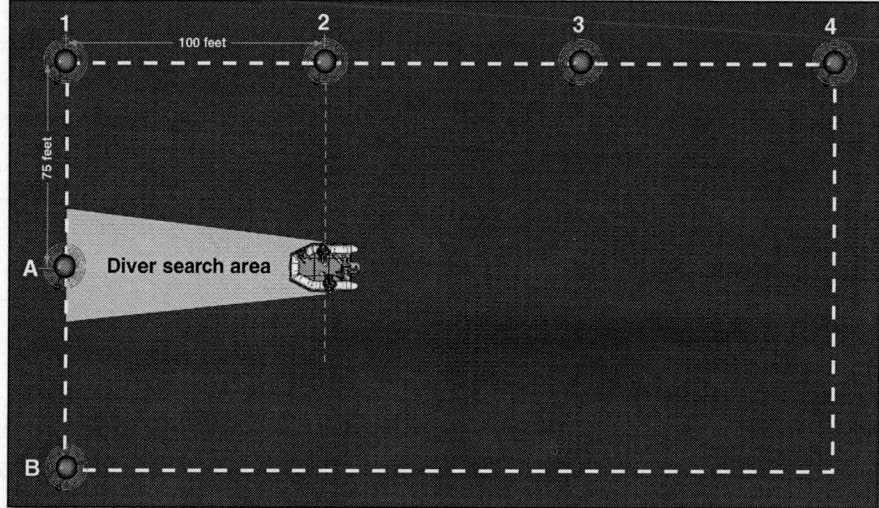

After searching along both sides of a boat, deploy a diver to search along the centerline of the previous positions.

vious search patterns. The tender should work the diver from the bow, sweeping him in a narrow path along the centerline from Buoy A to the boat's present position. Repeat this procedure until the entire block has been covered.

Diving From Small Boats

Certain safety rules apply to diving operations staged from small boats, some so obvious that they barely warrant mention in this sort of text. Boats should never operate under power in a dive area unless they're pulling a sled in a sled operation and there are no other divers in the water. If a boat does appear in a diver's area, the diver should be given a signal to stand by until the intruder has been removed. Also, teams should never engage in live boat operations. Rather, all operations should be staged from shore, docks, or stationary platforms. Divers should never be deployed when a boat is under way. Not only is it dangerous and unnecessary, it also prevents the profiler from determining the exact position where the diver went in.

When working in a small boat, all personnel need to stay low and close to the center of gravity. For that reason, divers should suit up onshore except for their fins, and possibly their full-face mask, both of which they should put on just before entering the water. Tenders should remain seated or kneeling, even when helping their divers in and out of the boat.

At least one person on a boat for a diving operation should be medically trained and equipped with a first aid kit. When operating in a rescue mode, that person will also be the one responsible for the initial care of the victim.

Generally, divers should deploy from the downstream side of the boat, not the stern, unless the stern is equipped with an actual dive platform. There is less room to work at the stern than along the gunwales, which consequently makes removal of the diver or victim from the water more difficult.

A diver deploying from along the gunwales of an inflatable.

Because of space limitations, you may wish to have the ninety-percent-ready diver waiting on a second boat no less than fifty feet away from the main boat. This second craft can also be used as a ferry, carrying personnel and equipment to and from shore.

If a boat is being used because there is a strong current, a weighted descent line at the stern may be required to help divers reach the bottom. A strong current will affect the ability of a tethered diver to reach the bottom in close proximity to an anchored platform.

Once the divers are in the water, the search patterns can be conducted as normal. On open water, you may find it necessary to use points on the boat, such as the bow and stern, to determine the turnaround points for each sweep. Remember, however, that divers affect the stability of a vessel or platform. If their search swings are too large, the stern of the vessel will travel in the same direction as the divers. This movement causes additional diver fatigue as well as a loss of pattern control. The three-point anchoring system will decrease the effects of this problem, but because of the length of the anchor scopes, even this method cannot entirely eliminate them.

Study Questions

1. Using a boat should be mandatory when the dive zone is more than what distance from shore?

2. In general, what is the best all-purpose type of boat for public safety diving operations?

3. What is the main drawback to using an airboat for public safety operations?

4. How is reverse achieved in a jet boat?

5. For public safety diving operations, if you have only one boat, how many people must it be able to carry?

6. Where can you find the rated maximum capacity of a boat?

7. True or false: When deploying an anchor, you should always toss it as far away from the boat as possible.

8. True or false: When using the hurricane anchoring system, there should ideally be an anchor 45 to 90 degrees off both sides of the bow and one off the stern.

9. After marking a large search area with buoys, where should the boat operate?

10. When searching a large-area grid, when should a diver first be deployed to search the area that has been under the boat?

11. When diving from a small boat, what pieces of gear may a diver put on while in the boat, as opposed to onshore?

Chapter 20

Law Enforcement Operations

Although unwitnessed drownings are supposed to be considered homicides until proved otherwise, responding personnel cannot help but be drawn to the idea that a given drowning was accidental. Popular culture trains us to view drownings this way. From the time that we were children, we have seen the drowning portrayed mainly as an accidental event, and as adults, we are trained in how to manage and prevent them in water safety courses.

Still, as experienced responders can attest, witnesses can lie, drownings can be homicides, and bodies can be dumped into the water to cover up foul play. Consider this scenario: A boat containing three young men overturns. Two of the men make it back to shore. One claims that he tried to save their friend but that he couldn't find him, and both of these witnesses are drunk. They're also both drenched. They claim that they were horsing around while they were fishing, and the boat tipped over. The dive team arrives, recovers the body, and the operation shuts down.

But wait a minute. If the same incident took place next to a bar, and if one man were found dead after supposedly falling down an alley cellarway, the scene would immediately be examined for evidence of possible foul play. Suppose the victim in each case has a head injury. Wouldn't law enforcement officers at the dive site want to know what caused it? Was it really that the victim hit his head on the gunwale as the boat capsized, or was he struck in the head with an oar or a beer bottle? Perhaps the victim owed money to one of the witnesses. Perhaps the victim was having an affair with the wife of one of the witnesses. Maybe they simply got into a drunken brawl. What, then, should a responding dive team do to ensure that a possible homicide investigation isn't ruined by rescue or recovery attempts?

Any dive team that will be involved in a recovery should train in how to handle evidence. Furthermore, any items or bodies found while operating in a recovery mode should be handled as evidence. If a team is searching for a body and the diver discovers a handgun, that gun could be important to a case, whether related to the body or not. Therefore, it should be marked and documented according to proper procedure, then entered into a chain of custody. Almost any drowning victim could have been the victim of a homicide, but that possibility may not even be raised until days or weeks after the body has been recovered. Thus, proper documentation and handling is imperative.

A significant percentage of dive teams are fire department teams, and these will respond to both rescue and recovery operations. Even though they may be excellent public safety dive teams, they usually have little or no training in handling evidence. A lack of training can have significant consequences. When proper information goes unrecorded, evidence is handled incorrectly, and a chain of custody is broken, an investigation is unlikely to result in a conviction. The same holds true when EMS personnel whisk away a drowning victim before the police arrive, having no understanding of the marks on the body, the foam in the airways, or the tiny red spots on the eyeballs.

Consider the case in which a boat operator, supposedly by accident, steered his craft over the water skier he'd been towing. The alibi was that the fallen skier was in a blind spot as the operator came round to pick him up. Although this sounds plausible, the investigating officers were skeptical of the story, based on the look of the wounds. They removed the motor and placed a mannequin against the propeller, running the prop in forward. A comparison of the slash marks on the mannequin to the wounds on the victim's body demonstrated that the victim was hit by a prop running in reverse, not forward. Further investigation revealed that the victim had been having an affair with the operator's wife and that the operator had purposely backed over the victim. The investigators in this case performed excellent police work. They didn't assume accidental causes, and they treated the boat as possible evidence. Sadly, the majority of drownings and other water-related incidents are treated as accidents, so the scenes are never thoroughly searched, and possible evidence is never put in the chain of custody.

Handling Underwater Evidence

Gathering evidence from a crime scene on land is a meticulous, time-consuming process designed to ensure that forensic scientists can learn as much from it as possible. Evidence recovered underwater should be treated no differently. Public safety divers, especially those on law enforcement dive teams, are often asked to recover objects from the bottom. With careful handling and treatment, it's possible for fingerprints to be retrieved from underwater items, even after years of submergence. It's vital to mark the location of evidence properly, and divers must be practiced at retrieving an item and placing it in a proper container, along with a sample of the sediment on which it was resting, all in conditions of zero visibility.

Documentation

A crime scene should always be documented thoroughly before any evidence is removed for preservation. To aid in this, attach a buoy to any evidence found

Placing a buoy on evidence found underwater allows a scene to be fully documented.

underwater so that measurements can be taken. Once the buoys are in place, accurate documentation of a scene can best be achieved through photography, or better still, with videography. Still, given black-water conditions and insufficient funding, proper debriefings with a narrative and profiling remain the most common representation of an underwater scene. A narrative is a complete descriptive account that encompasses information from the dive log and a court-ready profile map, along with the divers' interpretations of the whole operation. Once complete, the narrative is turned over to the investigator and becomes part of the case file. It isn't always necessary for each diver to write his own narrative, since a collaborative account may suffice, but this depends on the jurisdiction and the recommendation of the district attorney.

The profile map that results from the diving operation can be an important court document, showing how a recovery was made. Maps used in court should be large and in a variety of colors, drawn so that the jurors will readily understand it and be able to see how it relates to the case. Be sure that you clearly indicate the current at the time of the dive. Supplement that information with research from tidal charts and other sources, showing whether a current was running at the time the evidence might have entered the water. Information about currents is often neglected, but it is extremely important toward reconstructing what occurred at a crime scene.

When documenting a scene, you may need to use a combination of triangulation and direct measurement to attest to the location of the evidence recovered. Measure the distance from the object at the bottom to three fixed points onshore, preferably spaced far apart. The measurements must be of straight-line distances to the bottom, not to a surface position above the object. On the profile map, draw dashed lines between the location of the evidence and the permanent objects from which you measured, and indicate the distances. For reasons of depth and spatial relationships, the lines on the two-dimensional map will almost certainly be disproportional to the actual distances. However, if someone ever needed to return to the exact location, measuring off of the same three permanent objects would allow him to do so.

Photographs don't lie, nor do they make mistakes. They can be the most crucial evidence obtained in an underwater investigation. When possible, divers should take underwater photographs or video of all evidence before disturbing it. Evidence should also be photographed as soon as it is lifted out of the water, before it is taken from the site. Scene photographs should include an overview of the area, showing the body of water and anything else significant to the investigation, such as boats, docks, steps, gates, fences, or the shoreline.

A body submerged in cold, deep water for months with relatively little change can begin to decompose rapidly when it is brought up into warm air. Dive team personnel should seek instruction as to what to photograph, in the event that a coroner or other trained person isn't immediately available. Ideally, the body should be pho-

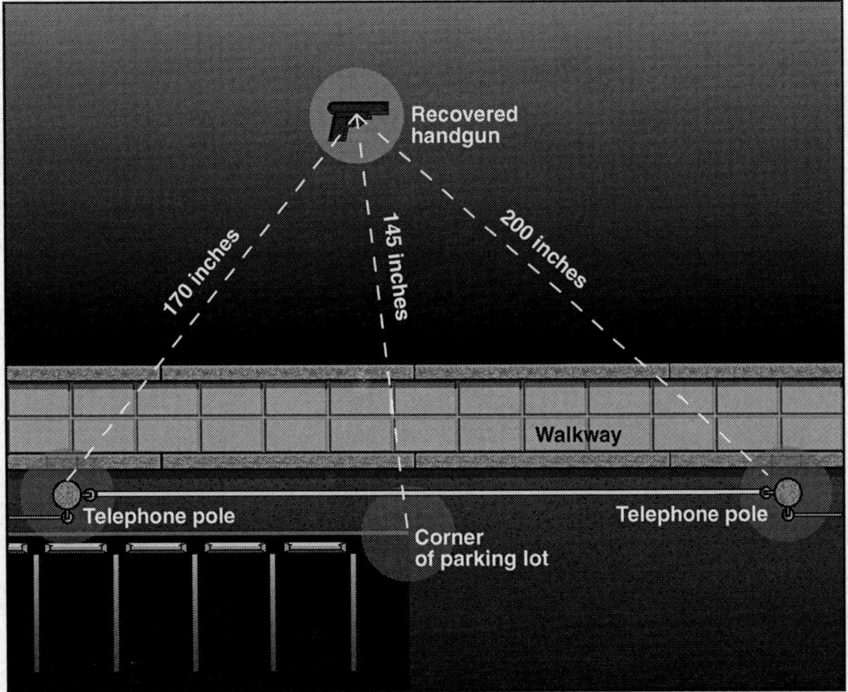

Measuring the distance from the object to three fixed points onshore can help you document a crime scene.

tographed by a coroner or other suitable examiner as soon as the body has been recovered, before changes take place.

Communications systems become extremely valuable during a criminal investigation, since they allow a diver to describe everything he sees and touches. They're even more valuable if the visibility is too poor for photography. Preferably, a tape recording will be made of the diver's entire time at the bottom.

Diving for Evidence

When dealing with submerged evidence, bear in mind that the crime scene may not be entirely at the bottom. For evidence to end up in the water, it had to get there from shore somehow. Thus, there may be other evidence nearby, on land. When approaching the incident scene, you should identify and preserve all possible points of evidence. This can be anything, from tire tracks to footprints to blood. Mark these items, and leave them undisturbed until a crime scene technician arrives. Also, take ground samples for forensic study, since they may correspond with clues in the clothing, skin, and under the fingernails of the victim.

Use the same caution in entering the water. Don't take the same path as the target you're looking for, if that path can be determined at all. Using a different approach angle will help preserve any evidence created on entry or dropped as the object passed through the water.

The actual dive operation for evidence should be a slow process. If the target objects are small, such as shell casings, divers will need to move at a painstaking rate, ensuring that they search every square inch of the bottom. They must also use extreme care when moving through the water itself, taking time to examine everything that they encounter. The bottom has to be searched by hands and fingers, not the whole body, as in searching for larger objects. Fragile evidence could easily be destroyed by a careless diver.

When searching this way, proper buoyancy is critical. Overweighted divers may accidentally push small items deep into the sediment as they drag across the bottom, or they may accidentally brush a small object into an area that has already been searched, causing it to be missed. Visibility in most cases is minimal to begin with, and stirring up silt doesn't help an investigation.

Containing Evidence

The key to retrieving submerged evidence is to begin by preserving it in the same water in which it has been resting. It should never be touched with bare hands, and it should be handled as little as possible. Every time it is manipulated, the surface of it will be disturbed, thereby disturbing potential clues. To prevent drying or oxidation, avoid keeping recovered items out of their fluid even for short times.

Some sources state that metal items, especially firearms, should be placed in kerosene on recovery. In practice, you should check beforehand with the local forensics laboratories to see how they want evidence to be handled in a particular case. Some forensic scientists prefer to work with evidence in the fluid in which it was found. If an item is scooped up into a proper container underwater, it should never make contact with air, thereby reducing problems stemming from oxidation.

If you are asked to place an item in kerosene for preservation, some sources suggest that you first place it in rubbing alcohol for fifteen minutes to displace any water from within it. After the alcohol bath, transfer the object to a container filled with kerosene. As always, check with your local laboratory's protocols beforehand.

When collecting the evidence, use rigid-sided containers. Plastic bags and other flexible containers will touch and rub against the item, marring fingerprints and other important markings. Using sealable plastic buckets to scoop up the object will automatically collect both water for preservation and samples of the bottom. For larger items, build containers out of PVC pipe. Objects underwater don't lose fingerprints for some time. Still, they must be kept in water to retain their value. If you

Use rigid-side containers to collect evidence and a sample of the bottom.

must handle an object to maneuver it into a container, take care to avoid touching any parts that are likely to bear fingerprints. For example, don't touch the slide of a semiautomatic handgun or the barrel of a shotgun.

Submerged Vehicles

If a vehicle is part of an underwater crime scene, the dive team has two options for searching it. The first is to seal it as well as possible, then tow or lift it onto land for the search. Following this method can damage the vehicle, however, especially if the lift isn't performed correctly. Also, if the doors, windows, and other openings are well sealed, water rushing from within may displace or destroy evidence. The shifting of the vehicle as it is dragged out will also move objects around. Although objects are likely to have been moved when the vehicle entered the water, it is still important to keep a crime scene as intact and reconstructible as possible.

The second option, then, is to search the vehicle underwater as thoroughly as possible. If visibility and the vehicle's position permit a clear view of the interior, divers should carefully document everything they see, whether by photography, videography, or real-time descriptions to a tender. Once this documentation is complete, the vehicle can be removed from the water. If the visibility is occluded, the vehicle should be carefully searched, preferably by opening only one door at a time. The location of any evidence the divers find should be carefully documented, then the evidence should be properly collected and taken to the surface. Of course, once the vehicle itself has been recovered from the water, it should be searched again.

In moving water, vehicles should be brought out before search, since the current may displace or remove any evidence found during the search. This is also important for reasons of safety.

Firearms

Handle guns with caution, since they may still be capable of firing, even underwater. During a training program in Washington, D.C., in 1987, a diver discovered a double-barreled shotgun with one hammer still cocked. On bringing it to the surface, it was discovered that this barrel still housed a viable 12-gauge shell.

Avoid mishandling evidence.

As mentioned above, avoid handling areas that are likely to contain fingerprints. You should also never insert any object into the barrel of a firearm. Doing so is likely to compromise the results or validity of a ballistics test. If possible, you should recover a firearm as you would any other piece of evidence, by collecting it in a sealed container with a sample of the bottom sediment included.

Crime Victims

If the team is operating in a rescue mode, a diver won't have time to waste taking pictures or giving detailed descriptions of what they find. He may only have a few seconds to determine the position a body is in before making his ascent.

If the operation is in a recovery mode, however, a diver will be able to perform an underwater examination. In conditions of poor visibility, this will have to be done by touch. The diver should work from the head toward the feet of the victim, reporting to the tender anything he discovers, such as jewelry, clothing, debris, and all else. Jewelry can easily be lost during the bagging procedure or the ascent, so if it is recorded now and doesn't make it to the surface, the team will know that a diver must be sent down to retrieve it. If the team lacks an underwater communications system, then the divers will have to practice gathering the information and forming a strong picture in their mind. Obviously, documentation by photograph or video is the best option, if visibility permits.

Once the initial underwater examination has been made, the body should be bagged for recovery. If it is impossible to bag the entire body at depth, at least bag and secure the hands, feet, and perhaps the head, making sure to cover the mouth and nostrils to preserve any evidence that could be trapped there. It's best to use a body bag with handles, as well as a fine mesh that will allow water to flow out but keep evidence in. A built-in lifting system is also of benefit.

If a victim is bound or weighted underwater, it is highly important to disturb the scene as little as possible. If underwater photography isn't possible, the victim and his bindings should be brought to the surface intact. Photographs need to be taken of all this evidence.

Preserving the Chain of Custody

Besides documentation and proper preservation, all items of evidence must also be tagged and have a documented chain of custody throughout the processing procedure.

Recording information with evidence is extremely important. Such information includes the case number; the location of the evidence at the scene; the depth at which it was found; a description of the item; the date and time it was located; the name of the diver who located the item; and to whom the evidence was released at the surface.

Note that this information should never be placed directly on evidentiary objects. To mark evidence, put a tag on its sealed container. Any marks made on evidence can ruin its value in court, since the evidence will have been changed in some way since it was found. Also, fingerprints or other important clues can be destroyed if someone marks the item directly.

It's vital to maintain a chain of custody for every piece of evidence. Otherwise, an item will be inadmissible in court. Each time that a piece of evidence changes hands, the receiver must sign a receipt for it. Although it will be necessary for evidence to change hands from a diver to a detective, or perhaps from a diver to a boat operator to a detective, try to keep any such chain as short as possible.

Recovering Underwater Explosives

No fire or police chief in his right mind would send anyone less than a qualified bomb disposal technician to retrieve a pipe bomb or hand grenade from inside a building, or anywhere else on land, for that matter. Still, dive team personnel with neither the necessary training nor the proper resources have been asked to recover blasting caps and other explosive devices underwater. Part of the problem is that, except for the military, there is no agency that has established, recognized teams with the expertise and equipment to conduct explosive-recovery operations underwater. Because of various legal issues, military personnel, such as Navy Explosive Ordnance Disposal (EOD) divers, aren't allowed to perform civil law enforcement duties. The exceptions to this rule include downed airliners, which involve federal mandates.

When an explosive goes off on land, the principle agents of injury are projectiles, heat, and concussive pressure. Underwater, the first two causes aren't of main concern. Water, being eight hundred times denser than air, slows down projectiles and absorbs great quantities of heat. The concussive shock wave is a far greater problem, however. Air is compressible, whereas water is not. Since the human body is largely composed of fluids that have the same density as water, any shock wave in the water will pass right through the body, rupturing hollow organs such as the lungs and bowels.

Normally, the body can withstand a maximum shock wave of 50 psi without sustaining injury. Death is likely with waves of greater than 300 psi. If a one-pound (0.454 kg) charge of explosive were detonated underwater, a submerged diver would have to be at least 72 yards (69 meters) away from it to experience a shock wave of less than 50 psi.

One way that EOD divers can protect themselves is by surrounding themselves with a layer of air, as is provided by a dry suit and a diving helmet. If the blast occurs close by, however, this will not prevent injury or death. One method of recovery is to tie a line of at least 200 feet (61 meters) to the explosive, and then to clear the water of boats and people. Next, a trained and appropriately dressed technician walks away from the water, dragging the ordnance to shore, keeping the full length of the line in between. One problem with this method is the potential for entanglement with debris on the bottom. To solve this problem, the explosive can be secured directly to a lift bag that is remotely filled from land prior to being pulled in. Once ashore, the device can be rendered safe by a qualified bomb disposal technician. If it's too dangerous to secure a line or lift bag to the explosive, it may be possible to blow it in place by detonating a charge in close proximity to it.

If your team has been asked to perform an underwater operation to recover an explosive, or if you happen to come across such a device while engaged in other operations, contact the bomb disposal unit that has jurisdiction in your area. If your team lacks the proper training and certification, your SOPs should clearly state so and that your personnel shall be exempt from performing such dives. Contact the Bureau of Alcohol, Tobacco, and Firearms (ATF), since this agency has qualified explosive enforcement officers. The ATF also maintains a list of law enforcement and fire department personnel who are qualified in these operations. If domestic or foreign terrorism is known or a possibility, immediately notify the Federal Bureau of Investigation.

Study Questions

1. Until proved otherwise, unwitnessed drownings should be considered ________.

2. Most drownings are treated as ________ by responding public safety personnel.

3. Proper documentation of an underwater crime scene can best be achieved through ________ or ________.

4. What is a narrative?

5. True or false: Evidence should always be photographed as soon as it is lifted out of the water.

6. Why should you take ground samples from along the shore of an underwater crime scene?

7. Why is proper buoyancy critical when searching for small objects?

8. Aside from the water in which it was resting, some forensics laboratories prefer that metal objects collected as evidence be preserved in ________.

9. When collecting evidence, why are rigid-sided, sealable containers preferable to plastic bags?

10. How should a vehicle be searched in moving water?

11. What is the best way to recover a firearm?

12. If it is impossible to bag an entire body at depth, how should the diver prepare the body for ascent?

13. What information must be recorded for each piece of evidence recovered from the bottom?

14. True or false: If the chain of custody for a piece of evidence is broken, that item will be inadmissible in court.

Chapter 21

Special Considerations and Technology

Public safety divers may be called on any hour of the day or night. The water may be clear or darker than pitch. The temperatures at the surface may be blazingly hot or frigid. Given the endless variety of potential accident scenarios and unknown factors lurking underwater, there is simply no way to predict exactly what a team will encounter at an incident site.

Underwater, there is no essential difference between the low visibility of murkiness and that of night. When the environment above the surface is suffering from impeded visibility, however, shore operations at once become more complex. The most important part of night-diving operations is to make sure that no one disappears in the darkness. Tethers become even more important to prevent divers from surfacing unnoticed or in dangerous areas, such as under a dock or in a boat lane. Night operations require that tenders and divers wear waterproof, submersible, and preferably *water-activated* flashers so that no one can be in the water without being easily seen. Flashers are especially important if a diver becomes disconnected from his tether line. Audible signaling devices are also important if the flashers fail or if someone surfaces out of a direct line of sight.

Another important aspect of night operations is to light the scene for all. If the tenders and divers are limited to flashlights only, there is a greater chance of an onshore mishap, including improper setup of equipment. Ideally, a scene should be lit by overhead lights from a light truck, trailer, or portable light poles, or by floodlights permanently mounted on the dive vehicle. If no overhead lighting is immediately available, you can use vehicle headlights to light a scene. Run the engines to keep the batteries charged. For inaccessible areas, you can use several battery-powered or oil-burning lanterns for general illumination.

Setting up scene lighting requires both planning and direction, so the incident commander may wish to appoint someone to be in charge of the process. As you set up the lighting for a scene, position the lights for maximum shore visibility, but try not to point them out toward the water. Although doing so would provide a better view for the shore crew, it would also blind divers who need to look toward shore to see their tenders. If the scene can't be illuminated from overhead, position the lights parallel to the water's edge. Have someone standing far to one side of the tenders shine a spotlight on the divers when they're on the surface and on their bubbles when they're below. Divers should not be allowed out on their tethers beyond the range of acceptable illumination. Otherwise, a diver could surface in distress without his tender realizing that there is a problem.

Because artificial light is no substitute for daylight, team members should also each have a small flashlight for close-up work, such as for checking gear. Divers should also have small, submersible flashlights, secured in a BCD pocket, that they can use for signaling, reading gauges, or any other purpose.

Boat operations become particularly difficult at night, because it is harder to see the shore and determine whether or not the vessel is holding in one place. Besides meeting Coast Guard regulations for lighting, all boats must be equipped with floodlights, either mounted on poles or on the radar bridge of a bigger craft.

Since the air temperatures are lower, tired personnel will tend to feel colder faster, so adequate thermal protection is doubly important. Personnel should move about more slowly to avoid tripping over shadowed ground. Divers may also find that they're more uneasy diving at night, simply because there isn't any daylight when they come up from below. Profilers may find it necessary to spotlight the opposite shore or deploy buoys to help mark the search boundaries.

Because of the added complexities posed by night operations, the time of day should be an important part of your risk-to-benefit analysis. If the operation is only a recovery, perhaps you should suspend it until daylight. Your SOGs should set parameters for support resources, and if those requirements are not met, the operation should not be attempted. Your SOGs should also state that if divers don't have the training and certification for night diving, then they shall not be allowed in the water for such missions. Practice at night is important, since you don't want your team's first night experience to be a real one.

Moving Water

If they can make the distance at all, the average public safety diver can barely swim 450 yards without fins in less than fifteen minutes. Making that swim in fifteen minutes requires a speed of 90 fpm, or 0.9 knots. If you add a mask and fins, the average public safety diver will struggle to complete a 900-yard swim in twenty minutes, a pace that is also less than one knot. Consider, then, how fast a diver is able to swim while wearing a scuba cylinder, pony bottle, BCD, and other gear, all while under the restrictions of an exposure suit and weight belt. A strong swimmer in full public safety diving gear may, at best, be able to swim at a top speed of one knot, but he'll soon be exhausted and won't be able to conduct a good search during such a swim.

Suppose that you show up at the scene of a body recovery where the water is moving at two knots. The dive coordinator and safety officer both decide that the current is too fast to warrant a recovery, so they deem it a no-go operation. Still, the team has nothing in its SOGs to that effect. Because you didn't dive, the family initiates a lawsuit. With nothing in the SOGs about moving water, how can the team back up its decisions in court? If for any reason your team refuses to dive in fast water, you should videotape a test of the current. This will at least help verify the circumstances at the incident site on the day in question.

In as little as one-half knot, shored-based diving operations are neither safe nor effective. Because divers sweep roughly parallel to the shore in such an operation, they are alternately swimming into the current or with it. That situation alone creates five significant problems. First, when the divers move downstream, they may be pushed along too quickly to make an effective search. Second, they'll exhaust themselves and their air supply on the return sweep upstream. Third, in turning around, they'll be broadside to the current, and thus more prone to being driven onto rocks or other obstacles. Fourth, they won't be able to maintain good search patterns. Fifth, even if a diver is able to swim well against the current, the moving water will bend his tether line so much that the search pattern is compromised.

Because of these five problems, your SOGs should state that if the current is over one-half knot, the operation must be staged from an anchored platform so that the divers can be deployed with the current and their search patterns can run perpendicular to it.

As the speed of the water increases, the problems associated with it also increase. Even working from a boat, currents of greater than 1½ knots can forestall operations. Because of its force, such a current can pin a diver against an obstacle, and he may be unable to move against it enough to free himself. Entanglements also become more difficult to manage. Water of 2½ knots can tear a diver's mask off his face if he turns broadside to the current, and as mentioned above, turning face-on to such a current can depress the purge button of a second stage. Moving water affects tenders as well, since a tender may be unable to fight the drag of any current of more than 1½ knots. A tender may not be able to bring a diver back toward the boat. Worse, there is a strong possibility that the tender will be pulled overboard by the diver's tether line.

Finally, consider what happens to a tethered diver in a current. Suppose that a diver is using a running-line pattern for his search area. He will be carried downstream some distance during the time it takes him to descend to the bottom. Once the diver reaches the end of the tether line, the current will lift him as the line tries to straighten out, much like a hair blowing in the wind. Thus, the length of each sweep on the bottom with be shortened, and once blown to the surface, the diver will have to be pulled or transported back to the dive boat.

Since divers cannot be tethered in fast-moving water, why not just let them dive untethered? In moving water, an untethered diver won't be able to stay within the search area, and it will be more difficult to reach him in the event of an emergency. If the water is moving at $2^1/_2$ knots (250 fpm) and a diver is in thirty feet of water, where should the pickup boat and backup diver be downstream? From a depth of thirty feet, a diver would be at the surface in about sixty seconds, given a nominal ascent rate of thirty feet per minute. If untethered, he would be no less than 250 feet downstream of his last known underwater location when he surfaces. It would take a safety boat a good thirty seconds after that to see the stray diver and retrieve him. Within that time, the diver would have traveled another 125 feet along the surface. Since the safety boat needs to come upstream to a diver in distress (rather than downstream, which would risk running over him), it should be stationed no less than 300 to 350 feet downstream of the farthest point that the tender intended to place the diver.

Because diving untethered in a strong current isn't an acceptable option, the best techniques for diving in water of more than one-half knot are the running-line search and the static search. A running-line search involves many up-and-down trips for the diver, presenting an increased risk of equalization problems and lung overexpansion injury. The better technique is the static-line platform search. For this technique, the diving platform itself is moved while the diver searches at the end of a tether of fixed length. In a river, the best method is to set up a static line system, in which a line is anchored from one shore to the other, and rigged so that a diving platform can be maneuvered upstream, downstream, and across the search area. If only one shore is accessible or if the distance between shores is more than two hundred feet, anchor the diving platform using the hurricane anchoring system.

Once the platform has been set up, divers must be let out on enough tether line so that they won't be pulled up from the bottom by the current. With the divers deployed, you'll have to determine at what distance from the platform they can comfortably lie on the bottom, facing upstream. That distance will be affected by both the depth of the stream and the strength of the current. Generally, the shallower the water, the farther out the diver must be to avoid being ripped off the bottom. However, tenders must also ensure that their divers stay close enough for a backup to reach them safely. Ideally, underwater communications systems are used for this type

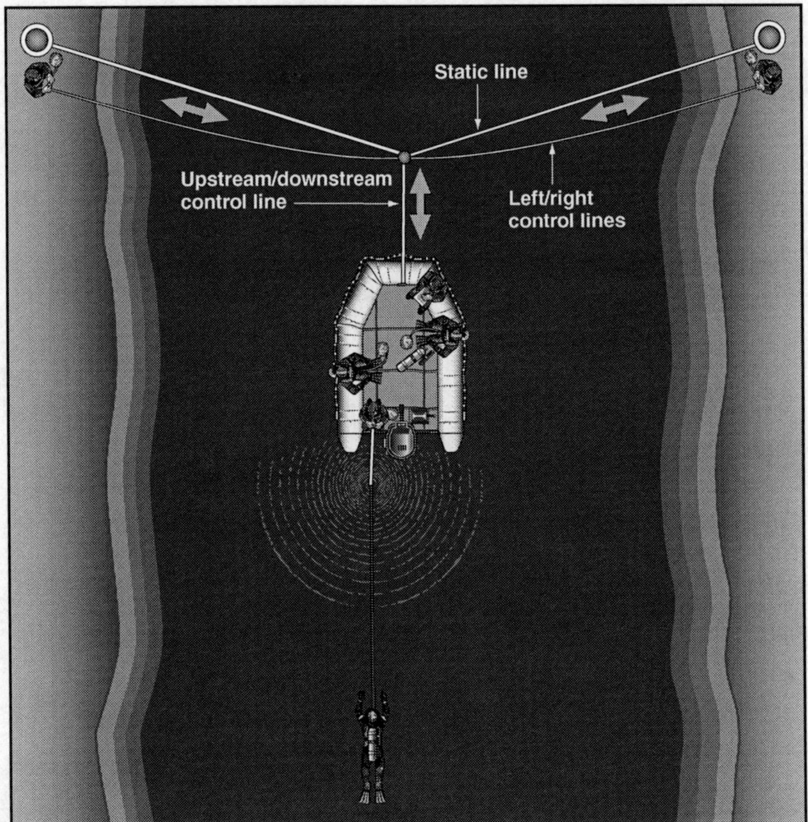

Static-line platform search.

of diving, although you should ensure that the tether isn't too long for line-pull signals, just in case the communications system fails.

Divers will have to add four to five more pounds of lead to their weight belts to help them stay on the bottom in a strong current. Once they're on the bottom and have searched the immediate area, the tender signals the diver, and the entire platform is pulled a few feet to the right or left. The diver again searches the immediate area. Once the span of the search area has been covered, the platform is lowered two or three feet downstream, and the diver begins another sweep. The advantage of working this way is that the diver doesn't have to fight the current, and he is always at a fixed distance from the boat.

This method is accompanied by a number of problems. Adding several pounds to the weight belt is in itself a safety risk. For this type of operation to be successful, there must be little or no bottom debris, since the shallow angle of the tether line and the movement of the platform increase the risk of entanglement. Contingency plans are also more difficult, for if the backup diver is simply deployed down the tether, the current will push him into the primary. If the backup diver pulls the primary's line, the primary may be lifted off the bottom, worsening the entanglement.

Put simply, diving operations in moving water are logistically difficult and can be quite dangerous. Your SOGs should therefore state that unless everyone involved in an operation has had the proper training, public safety diving will not be done in water moving faster than $1^1/_2$ knots.

Ice Diving

If public safety diving is a professional form of technical diving, then public safety ice diving is its advanced form. Ice operations present both an overhead environment and extreme cold. Tenders must be trained in how to control a diver in an overhead environment with a single exit, and no primary diver should be let out more than one hundred feet from the point of entry. This will mean that a diver can be located more quickly in the event of a disconnected tether. Even if more than one hole is cut, ice diving really only offers one way in and one way out. Although public safety diving is exempt from OSHA standards, confined-space

Ice operations present both an overhead environment and extreme cold.

Diving under an overhead environment such as ice requires thorough training and contingency planning. (*Photo courtesy of Pete Nawrocky.*)

operations are not. If you consider ice dives to be confined-space dives, you further realize the importance of using one diver down, two backup divers, and redundant air supplies.

Even without ice, diving in water of 45°F or below should be treated as ice diving in terms of thermal protection and the amount of time that divers can spend underwater. Fifteen minutes should be the limit for a basic dive, with a possible five-minute extension. Extreme cold can cause regulators to freeze and free-flow, which quickly results in an empty air tank. Worse, one minute of attempting to control a second-stage free flow in frigid water can make your mouth so cold that you cannot hold your mouthpiece in place, and the tip of your tongue can literally freeze. Naturally, immersion in frigid water only accentuates the dangers of hypothermia, a risk that is present in all diving operations. Ice diving demands that divers wear dry suits with adequately insulative undergarments. Although some teams wear wet suits, these are far less effective against the cold. A diver might be able to tolerate a fifteen-minute dive in a wet suit, but what happens when he becomes entangled on the bottom for forty-five minutes? Even if he is freed, there is a good chance that he will be incapacitated due to hypothermia. Also, divers who wear wet suits may not be able to make repetitive dives in cold conditions. Cold can wipe out entire rotations of

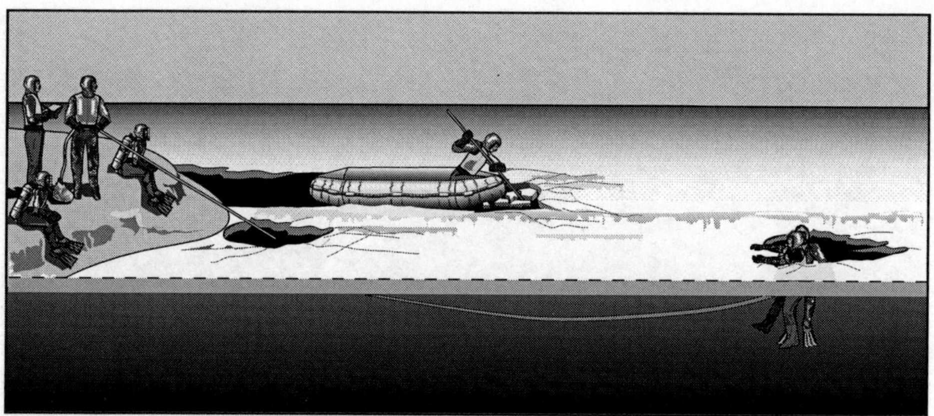

A diver deployed under the ice can help maintain a victim's buoyancy, buying time for a surface rescuer to reach the site.

divers. By the time the primary has completed a fifteen-minute dive, the backup and the ninety-percent-ready diver may be so cold that they will be unable to go below. Cold also increases air consumption underwater.

The ice itself presents problems, since no ice may be considered safe. In responding to an incident in which a person has fallen through, it has already been proved that the ice is untenable. In situations where rescuers are unable to reach a victim on the surface, a diver can be deployed below the ice at a point where the rescuers are falling through. Coming from below, the diver can create buoyancy for the victim, allowing the surface rescuers more time to destroy ice as necessary and bring a viable platform to the location.

On the positive side, victims who plunge through the ice, experiencing sudden immersion in cold water and quick drowning, stand a relatively good chance of being successfully revived. Thus, the potential benefits of an ice diving operation are high.

Heavy Weeds

Aquatic plants of all varieties can make underwater searches more difficult. They snag valves and fins, and they entangle divers. As a tether line sweeps back and forth, it can pick up so many weeds that the diver will struggle just to overcome the drag. Worse, search patterns can be ruined as the weeds cause the line to bend.

When searching for a body in weeds, many divers burrow their way through the weeds to the bottom and then begin searching along as if they were in a heavy jungle. That approach has a major flaw: If the diver had to dig his way to the bottom, how would the body have gotten through? In reality, bodies land in the upper

Heavy weeds can entangle divers and weigh down tether lines.

Search for a body along the tops of the weeds, not in them.

portions of dense aquatic weeds and stay there. Instead of plowing through the weeds, a diver should swim just over the tops of them, searching with his hands. Often a diver will feel a depression caused by the weight of the body before finding the body itself.

Searching for small, heavy objects in the weeds, such as weapons, is more challenging. If the object is metal, consider using an underwater metal detector. Another option is to wait to conduct the search during the winter months, when water weeds tend to lie flatter or may even have died. A final option is to cut and remove the weeds to search an area. Obtain permission for this, if necessary.

Since tether lines generally follow the slope of the bottom as it drops away from shore, weeds along the bottom can entangle the entire length of the line. By shifting to a boat operation, working the diver from the shore and toward the boat, the tether line will be angled up and away from the weeds, decreasing the amount of interference. The steeper angle of the tether will cause less conflict with the weeds and allow for an easier, more accurate search.

Altitude Dives

Atmospheres of pressure underwater are based on atmospheric pressure. A lower atmospheric pressure means that a diver need not go as deep to encounter water equal to one atmosphere. For example, if the atmospheric pressure at a high-altitude dive site is 10.5 psi, divers will encounter a pressure of two atmospheres at about twenty-five feet, rather than at thirty-three feet, as at sea level. Thus, a depth of twenty-five feet at this high-altitude site is equivalent to making a thirty-three-foot ocean dive.

Standard dive tables and computers are calibrated for sea level. When you dive at higher altitudes or move to a higher altitude after diving, however, your body is returned to a lower atmospheric pressure than is accounted for by the tables or computer. This creates a higher differential between the pressure of the nitrogen still dissolved in your body tissues and the ambient pressure.

Suppose a call goes out for a drowning in a mountain lake at an elevation of 8,000 feet MSL, where the atmospheric pressure is 10.92 psi, or 0.742 atmospheres (ATM). In addition to changes in depth gauge readings and buoyancy, and assuming that the divers have had at least twelve hours to adjust to the local pressure, the length of time that they can spend underwater at that altitude will be decreased. A dive to thirty feet will be equivalent to a dive in forty feet of ocean water. If a diver uses regular dive tables rather than altitude tables, he will be at serious risk of decompression sickness. High-altitude dives also require slower ascent rates.

For any dives above 1,000 feet MSL, you'll need special dive tables and specific training.

Sleds and Tow Bars

Given clear water conditions, sleds and tow bars can be used to search large areas quickly. Drawn behind a motorboat, one or two divers can cruise above the bottom at a reasonable rate, scanning on both sides for signs of the target. Sleds and tow bars are excellent for water in which the visibility is greater than ten to fifteen feet. Sleds look like underwater toboggans, and some even have airplanelike controls to facilitate ascents and descents. A tow bar is simply a bar or yoke handle that a diver holds by hand. The quality of the search depends on the operator of the tow boat, who must be able to follow close reciprocal courses to maintain even, overlapping sweeps of the search area.

Using these devices creates a tremendous amount of drag on the tow boat. Thus, any setup designed to pull a sled or tow bar must be rigged so that the line is centered on the stern of the boat. Otherwise, the boat will tend to veer off course. Use a ski harness, or rig a Y-shaped yoke from webbing or strong line, then connect it to both sides of the stern.

To ensure that the tow boat follows a predictable course, buoy the lines of the search pattern with floats about fifty feet apart. Place different-colored buoys at the ends to signify the turnaround point. Remember that the distance to the turnaround points is governed by the length of the tow line. The diver trails a long ways behind the tow boat, and he won't have searched an entire leg of the pattern until the boat is far past the end of the leg.

Tow bar.

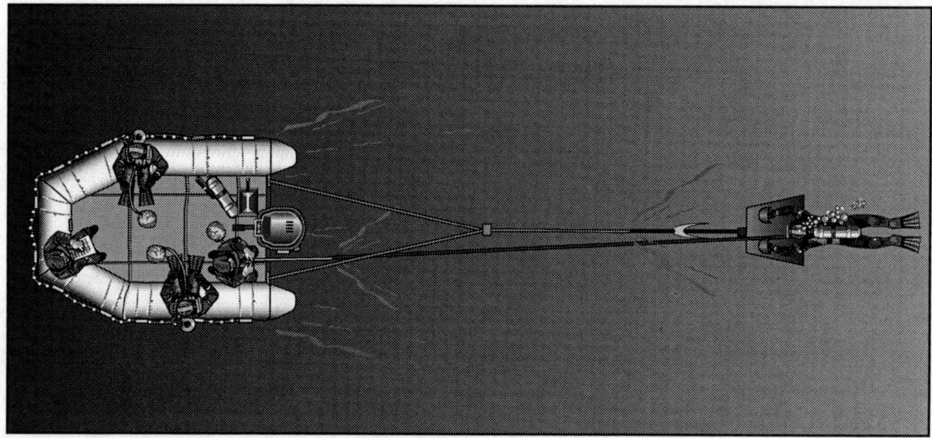

Any setup designed to pull a sled or tow bar must be rigged so that the line is centered on the stern of the boat.

During the turnaround, the diver will simply land on the bottom and wait as the tow line goes slack. The tender at the stern of the boat will need to handle the lines to ensure that they don't become caught in the intake or propeller. When towing divers, the boat's speed will depend on the type and size of the sled or tow bar, the number of divers being pulled, and the visibility of the water. In no case should it go faster than $2^1/_2$ knots. Faster speeds increase diver heat loss; put too much pressure on regulator diaphragms, causing them to free flow; and flood or possibly tear off a mask. Divers may wish to use side-breathing regulators, or they can build a small shield to attach to the face of the second stage to prevent the flow of water from depressing the purge button. Be sure that such a shield will still allow you to reach the button yourself.

Although some sleds come with signaling devices, many are very simple and homemade. To facilitate communications with the tow boat, attach small sections of rubber hose or plastic pipe every five to ten feet along the tow line. Then, run a secondary line through them, to the tender. That line can be used for line-pull signals. In another variation, run the cable of a hardwired communications system instead. When using the former, you'll have to work out some new signals, indicating such commands as "Prepare to be towed," Prepare to turn around," "Speed up," "Slow down," and the like.

When flying a sled or tow bar, divers should maintain as high an altitude above the bottom as visibility will allow. On finding the target, he can let go of the sled, drop a buoy, and either recovery the object or signal the boat.

Although useful search tools, sleds and tow bars present a few hazards to those who use them. Divers should never be tethered to them, because of the chance of entanglement and strangulation. Divers can also suddenly find themselves in areas of low visibility, posing a danger of crashing into unseen obstacles. If the visibility drops off, divers should release the sled or tow bar and make a normal, slow ascent to the surface. Whenever possible, use two divers on a sled, so that one can be on the lookout for obstructions while the other one searches. Because they're moving at a fairly high rate through the water, divers face a greater risk of cold stress and hypothermia. Proper apparel is mandatory. Perhaps the most serious risk is that of decompression sickness and lung overexpansion injury, since the unwary diver may easily make altitude changes at too high a rate. Any ascents and descents over undulating terrain must be made slowly enough to equalize your ears, mask, and sinuses. At the end of such an operation, never ride the sled to the surface. Instead, simply release the sled and make a normal ascent.

When conducting sled operations, have a backup diver and tender on board the support vessel, as well as a ninety-percent-ready diver.

Rebreathers

A rebreather is an underwater breathing apparatus that removes carbon dioxide from a diver's exhalations, adds a small amount of gas, and then recirculates the mixture for the diver to breathe again. A closed-circuit rebreather can eliminate exhaust bubbles and provide a diver extremely long bottom times.

Rebreathers have been in use for decades. In fact, pure-oxygen rebreathers were used by frogmen in World War II. Only recently have they become more available to and affordable for the civilian diving community. Many public safety divers have suggested using rebreathers, noting the greatly extended bottom times that they allow. However, except for tactical law enforcement teams that may need the no-bubble advantage to run covert operations, rebreathers are of little use to public safety teams. One reason is that public safety divers don't benefit from longer bottom times. Attention span, not depth or air consumption, is generally the limiting factor. In low-visibility water, divers generally cannot search effectively for longer than twenty or twenty-five minutes. Although some may argue that the extended gas supply of a rebreather would benefit a diver trapped on the bottom, there would still need to be some type of fully redundant system to act as a backup in case of rebreather failure.

Rebreathers also require extensive training and underwater experience. A rebreather must be properly set up before each dive, a skill that must be practiced frequently. More delicate and maintenance intensive than standard scuba gear, a rebreather requires a lengthy setup procedure before a dive, and many aren't one hundred percent reliable. The preparation that they require makes them unsuitable for the rapid deployment modes of public safety diving. Finally, the cost of a rebreather is also prohibitive, since they range from about $3,000 for a basic, semiclosed-circuit unit to $14,000 for a fully closed one. Although they may have applications for some, rebreathers are too risky, too expensive, too impractical, and simply unnecessary for the vast majority of public safety teams.

Electronic Navigation

Electronic navigational systems aren't new. Loran (Long-Range Navigation), a radio-navigation system using a network of land-based transmitters, has evolved through several stages since World War II. Developed for marine use along the coasts and in the Great Lakes, loran has found application among recreational boaters, as well as in the national airspace system for both supplemental en route and nonprecision approach operations. An onboard receiver compares the discrepancy between the signal from a master station against the sequenced signals from at least two other stations in a chain of secondary transmitters. The U.S. Coast Guard operates twenty-seven domestic transmitters divided into eight chains, a system that is compatible with and supported by a similar Canadian system. Measuring the time difference of the signals down to a discrepancy of 0.1 microseconds, loran is able to provide position accuracy to within approximately one hundred feet. Because the system operates in a low-frequency band of around 100 kHz, its signals are prone to distortion by the signals of adjacent bands. Thus, all loran receivers must be equipped with selective internal filters to limit interference. Loran is also affected by terrain, making it ineffective for inland waterways.

A more recent development, the Global Positioning System (GPS), is based on an array of twenty-four satellites, each of which transmits a unique signal. As with loran, a GPS receiver must acquire at least three signals to provide a two-dimensional fix of latitude and longitude. The acquisition of four signals can provide elevation information. GPS can also be used to plot a course to a specific location, or waypoint. In theory, the system should be capable of placing a user within a few feet of a specific location anywhere in the world.

Side-Scan Sonar

Another tool that is gaining use in underwater public safety operations is side-scan sonar. Although this technology has been in use for decades, it has previously been unable to detect objects as small or as soft as a human body. Improvements in the resolution of images mean that side-scan sonar may be the way of the future for rescue/recovery operations.

One of the great advantages of this technology is that visibility has no effect on how well the sonar device sees its targets. Instead, side-scan sonar creates visual images of objects on the bottom by using sound waves projected from either side of

Life-size mannequin at a depth of thirty feet in black water using side-scan sonar. (*Photo courtesy of Marine Sonic Technology Ltd.*)

an emitting unit, or tow fish, pulled behind a boat moving at slow speed. The sonar unit then receives the reflected sound waves and uses the pattern and timing of them to create an onscreen visual image. The denser an object is, the more sound it will reflect, making its onscreen image brighter and more definite. Subtle differences in the length of time that it takes to receive a return signal allow the unit to determine the shape of an object. A portable computer stores and processes those images, allowing an operator to review them.

The sonar display produces an image of the bottom as seen from above. The contour of an object will appear well defined and lit on one side and shadowed on the side that is blind to the tow fish. Besides simply creating an image, some side-scan units can measure lengths, areas, and heights above the bottom for entire objects or even parts of them, thereby helping the user determine whether or not a given target could be the correct size for a human body. Height above the bottom isn't measured directly by the return, but rather, indirectly, based on the length of the object's sonar shadow and the angle to the top of the return. The computer uses trigonometry to solve for the height of the object.

Side-scan sonar has two major applications for public safety operations. First, this technology can allow you to conduct a rapid search of a large area. Given ideal conditions on the bottom and skilled operators, an area of one million square feet can be searched in a matter of hours. The second application is in searching deep water. Since operating in limited-visibility water at a depth beyond sixty feet is impractical and risky for public responders, and since there is no chance for a successful rescue beyond such a depth, side-scan sonar can be used to locate a target without risking a diver's safety.

Among its many drawbacks, side-scan sonar requires that the tow boat be piloted along steady, straight courses. It cannot be used if the bottom is strewn with debris, since the shape of a human body could easily pass unnoticed among a host of irregular shapes. Debris may also snag the tow fish. The greatest drawback to this technology, however, is its cost. A unit may run upward of thirty thousand dollars. Such an expense may seem more reasonable when one considers the cost of mounting a full operation in a paid department. If your team decides to purchase such a unit, be sure to buy the top-of-the-line model, since less expensive ones don't offer the resolution necessary to identify a human body.

Handheld Sonar

Small, handheld, battery-powered sonar systems are also available. Unlike side-scan sonar, most of these devices don't create a high-resolution image. Rather, they indicate that an object is present by providing a diver with either audible signals via headphones, or a visual display.

Generally, handheld sonar units have two modes. One mode is passive, in which the unit functions only as a receiver capable of locating underwater acoustic beacons. In the active mode, the unit sends out an acoustic ping, and the return signal is then electronically interpreted for either an aural tone or a visual display.

To locate an object using handheld sonar, a diver starts on a known bearing and makes a slow, full circle, keeping the device pointed outward. If he receives a return signal, he notes the compass heading and approximate range of the signal. He may then either swim toward the object for investigation or surface to report his findings to the support crew.

If no object is found in that full-circle sweep, the diver should signal the surface personnel, then move to a new position to commence another search. As always, the search patterns should overlap to some degree. Repositioning the diver is best done by tether from a stable, fixed point, as from a vessel moored on a three-point system.

Handheld sonar can be extremely useful in searching a flat bottom for larger objects, such as boats, aircraft, and vehicles; thus, they can be of value for teams that work large lakes or flat ocean bottom, especially near airports. If the underwater terrain is too rough, has too much debris, or is laden with kelp, handheld sonar will be markedly less effective. Certain models don't operate on a high-enough frequency and have proved limited in their ability to locate a human body underwater.

Metal Detectors and Magnetometers

Metal detectors work by transmitting radio waves, which are reflected by metal objects. A receiving circuit amplifies the reflected waves to signal the user. A metal detector will detect any type of metal, which makes them useful for general searches, as at a crime scene. However, such a broad range of detection can be problematic, since they will also locate aluminum cans, bottle caps, belt buckles, discarded wheels, and other junk.

Magnetometers are available in both handheld and towable versions. Unlike standard metal detectors, magnetometers detect only ferrous objects. Thus, they won't lead a user off on a search for every scrap of metal present, and they work well for detecting almost any kind of metallic weapon, such as guns and knives. However, because they are limited in scope, they won't allow a user to find objects that might also be important, such as brass shell casings or lead bullets. Similarly, a towed magnetometer may be of limited use in searching for an aircraft, since most aircraft are principally composed of aluminum.

Remotely Operated Vehicles

A remotely operated vehicle (ROV, sometimes called a rover) is an unmanned robot propelled through the water by small thrusters. ROVs are typically equipped with one or more cameras, and they may have robotic arms as well. Usually they are powered and controlled by means of a tether leading to an operator at the surface.

Remotely operated vehicle, or ROV.

Although ROVs can conduct search operations without having to jeopardize human life, they do have numerous disadvantages. Even to purchase, they can be cost-prohibitive for a team. Because they have numerous intricate and often delicate parts, they require a high level of maintenance and care by skilled technicians. Like side-scan sonar, they also require skilled operators. Unless they are outfitted with an expensive sonar system, ROVs are limited to water with good visibility, since the ROVs pilot relies on its cameras to see what is in close proximity to the machine.

Although ROVs can have some applications for search and recovery operations, they are impractical for most public safety dive teams.

Study Questions

1. Ideally, how should a dive site be lit?

2. As long as a diver is tethered, should he be allowed beyond the acceptable range of illumination?

3. True or false: A team's SOGs should stipulate that a current of as little as one-half knot will preclude public safety diving operations staged from shore.

4. What is a static-line platform search?

5. True or false: Generally, when using a static-line search, divers must be deployed farther out in deep water to avoid being ripped off the bottom by the current.

6. Approximately how much extra lead might an average diver expect to need to help him stay on the bottom in a strong current?

7. In terms of thermal protection and the amount of time that a diver can spend below, an operation in water of ________ or less should be treated as an ice diving operation.

8. True or false: Cold increases air consumption underwater.

9. Where should a diver first search for a missing body in an area of dense weeds?

10. Starting from standard conditions at sea level, a diver will encounter a pressure equivalent to two atmospheres at a depth of approximately ________.

11. For any dives above an altitude of ________, you need special dive tables and specific training.

12. Sleds and tow bars should be connected to the tow boat by means of a ________ or ________.

13. In no case should a tow boat pull a sled faster than _______.

14. What is perhaps the most serious risk of using sleds and towbars?

15. As an argument against using rebreathers, what is generally the limiting factor for divers in public safety operations?

16. What government entity oversees and operates the loran network?

17. Loran provides position accuracy to within approximately ________.

18. To obtain both position and elevation information, a GPS receiver must first acquire the signals of how many satellites in the network?

19. The practice of randomly altering GPS signals every few minutes to induce a small amount of error in position information is known as ________.

20. In correcting for GPS satellite error, the Differential Global Positioning System is able to provide a navigational fix to within what distance?

21. What sort of visual image of a target object does side-scan sonar produce?

22. What two major applications does side-scan sonar have for public safety operations?

23. What are the two modes of handheld sonar?

24. What is the principal limitation of metal detectors?

Appendix

Calculating Surface Air Consumption Rates

It is important that surface air consumption rates (SACR) be calculated and recorded for each diver at the end of each dive. Eventually, each diver will know his average SACR for different kinds of dives with different types of equipment, such as with standard or full-face masks. This information can be used to help determine whether a diver is capable of making a particular dive; whether a diver can have a five-minute extension; and whether a diver may have experienced a problem, as evidenced by an unusual SACR.

The reason SACR is calculated rather than the depth air consumption rate (DACR) is because a known average SACR can be extrapolated to a DACR for any particular dive. For example, a diver with a 20-psi/min. SACR will use 29 psi/min. at fifteen feet, 40 psi/min. at thirty-three feet, and 60 psi/min. at sixty-six feet.

It's possible to calculate the SACR in terms of psi/min. However, since any organization will rely on cylinders of various sizes, including pony bottles, SACR must instead be calculated in terms of volume. A reading of 25 psi in a 100-ft.^{3} cylinder, for example, represents twice as much air as 25 psi in a 50-ft.^{3} cylinder.

The first key to finding your SACR is your logbook. You should record your starting tank pressure, ending tank pressure, the depth, bottom time, and the size of the tank that you used. Suppose that you began a dive with 3,000 psi and ended with 1,100 psi. You went to twenty-five feet for a total of twenty minutes with an 80 ft.^{3} tank.

First, determine the amount of tank pressure that you used by subtracting the end pressure from the start pressure.

$$\begin{array}{r} 3{,}000 \text{ psi (start)} \\ -1{,}000 \text{ psi (end)} \\ \hline 1{,}900 \text{ psi consumed at depth} \end{array}$$

By dividing the psi consumed by the length of the dive in minutes, you can determine how many psi you used each minute during the dive. The result, inclusive of descent and ascent times, is valid only for that depth.

$$\frac{1{,}900 \text{ psi consumed at depth}}{20 \text{ minutes (total length of time)}} = 95 \text{ psi/min. consumed at depth}$$

The increase of pressure at depth causes a diver to use more air from a tank than he would at the surface. Since you are trying to find your SACR, you must determine how much air you would have used had you breathed for the same amount of time at the surface. To do so, you must first calculate your depth in terms of absolute pressure, or ATA (atmospheres absolute).

$$\frac{\text{25 feet (depth in feet)}}{\text{33 feet (1 atmosphere, or ATM)}} + 1 \text{ ATM} = \text{ATA}$$

$$0.757 + 1 = \text{ATA}$$
$$1.757 = \text{ATA}$$

By dividing the psi/min. consumed at depth by the ATA, you can find how much air you would use per minute if you were to breathe for the same amount of time at the surface.

$$\frac{\text{95 psi/min consumed at depth}}{\text{1.757 ATA}} = \text{54.0 psi/min. consumed at surface}$$

Although these equations solve for the amount of tank pressure depleted, we still have not calculated the actual volume of air that you used. To find this figure, we must convert the surface psi/min. to cubic feet of air used at the surface. To do so, we must first find out how many cubic feet are used for each psi, thereby to arrive at a conversion factor for that particular tank.

$$\frac{\text{Tank size in ft.}^3}{\text{Tank's working presure in psi}} = \text{conversion factor}$$

$$\frac{\text{80 ft.}^3}{\text{3,000 psi}} = \text{0.027 ft.}^3$$

The next step is to find out how many cubic feet were used each minute. Anytime you need to convert psi to ft.^{3}, simply multiply it by the conversion factor.

54.0 psi/min. consumed at surface x 0.027 ft.^{3} = 1.458 ft.^{3}/min. consumed at surface

Your SACR by these calculations is 1.458 $ft.^3$/min. You would consume 29.16 $ft.^3$ in twenty minutes at the surface.

You can use the SACR to determine how long you would be able to breathe from a given cylinder at the surface, as well as at depth. To calculate the potential at the surface in minutes, divide the capacity of the tank by the SACR. Given an 18 $ft.^3$ pony bottle and the SACR calculated above, the result works out as follows.

$$\frac{18\ \text{ft.}^3}{1.458\ \text{ft.}^3} = 12.3\ \text{minutes}$$

To calculate how long a tank will last at a given depth, divide the result of the preceding equation by the ATA.

$$\frac{12.3\ \text{min}}{1.757\ \text{ATA}} = 8.5\ \text{minutes}$$

At a 1.46 $ft.^3$/min. SACR, an 18 $ft.^3$ pony bottle will last you 8.5 minutes at a depth of twenty-five feet.

Keep in mind that SACR is a highly personal number. Even given the same depth and conditions, air consumption may vary widely between individuals due to body size, cardiovascular health, lung capacity, and other factors. Because of these variables, you should keep track of your own SACR. It's also important to remember that breathing rate may vary from dive to dive, depending on the circumstances. A public safety diver working an actual emergency may be pumped up and breathing more rapidly from exertion than he would be at the same depth during training. An entangled diver breathing from a pony bottle may be using even more air.

Calculating the SACR is one of the most important aspects of dive planning, since it can give you an excellent idea as to how long a dive will last. However, because you can use more air than you predicted, you shouldn't rely on mathematics alone. Use SACR with caution, and always watch your gauges.

Sample Lifting Problem

Using the attached-buoyancy lift method, suppose that you want to lift a vehicle that weighs 2,894 pounds and that the vehicle is submerged in 50 feet of seawater.

The vehicle itself displaces 14 $ft.^3$, which represents 896 pounds of upward force (14 $ft.^3$ X 64 lbs./$ft.^3$ of seawater = 896 lbs. displacement).

Dry weight:	2,894 lbs.
Displacement weight:	–896 lbs.
Wet weight:	1,998 lbs

You will need 1,998 pounds of lift to bring this vehicle to the surface.

Since each cubic foot of salt water weighs 64 pounds, each cubic foot of air *at the surface* will displace 64 pounds, and therefore have a lift of 64 pounds. Dividing the 1,998-pound wet weight by 64 lbs./$ft.^3$ reveals that 31.2 $ft.^3$ of air is required to make the vehicle neutral in the water *at the surface.*

Since the pressure at 33 feet equals one Atmosphere Absolute (ATA), the pressure where the vehicle is located is equivalent to 2.5 ATA, as shown in the following equation.

$$\frac{\text{Depth}}{33\text{ ft.}} + 1\text{ ATM} = \text{Pressure at depth}$$

$$\frac{50\text{ ft.}}{30\text{ ft.}} + 1\text{ ATM} = 2.5\text{ ATM}$$

The amount of air needed to make the vehicle neutrally buoyant at a depth of 50 feet is 2.5 times that amount required at the surface.

$$2.5\text{ ATM} \times 31.2\text{ ft.}^3 = 78\text{ ft.}^3$$

Thus, 78 cubic feet of air is required to make this vehicle neutrally buoyant at a depth of 50 feet in seawater.

Once the proper amount of air has been added to make the object neutral underwater, you can begin lifting procedures. Ideally, divers would use a combination of lift bags that would hold no more air when full than is necessary to raise the object. Thus, any expanding air would simply vent from the lift bags. Alternatively, divers would ascend with the object being lifted so that they can vent air from the lift bags during the journey upward. The problem with this option, however, is that divers can rarely control lift bags well enough to maintain a maximum ascent rate of one foot every two seconds. If you are adding air to a lift bag and its rate of ascent exceeds that speed limit, you should get clear of the object. If the lifted object proceeds to the surface ahead of you, there is always a chance that rigging will fail or that the lift bags will spill their air on reaching the surface. In either of these scenarios, the object will plummet to the bottom.

Awareness-Level Duties

1. Know where to go to assess the scene and stage the operation.
2. Don PFD, gloves, and necessary PPE.
3. Perform rapid scene assessment.
4. Identify life and health hazards.
5. Assess the number and status of victims.
6. Establish communication with the victims.
7. Mark a spot on shore in front of each victim, if possible.
8. Implement the IMS. Establish a command post and an IC; establish staging areas, staging officers, safety officers, and other personnel and sectors as required by the size and complexity of the incident.
9. Call for the appropriate agencies, including the dive team, EMS, and law enforcement.
10. Direct arriving vehicles to form wind blocks and leave exit routes.
11. Secure and manage the scene; determine and mark the hot, warm, and cold zones.
12. Secure witnesses for interviewing by law enforcement.
13. Draw a profile map of the area and the most likely victim location. Find a safe place to deploy divers.
14. Reassess the status of the victims.
15. If higher-trained personnel haven't yet arrived, begin interviewing witnesses and document information on the profile map.

Dive Site Checklist

1. Incident command profile map board for documenting which areas have and have not been searched.
2. Witness-interviewing paperwork.
3. Profile and record-keeping paperwork.
4. Accountability tags or board.
5. Name, telephone number, and next-of-kin contact information for all divers.
6. Tarps.
7. Fire or police tape.
8. Evidence recovery containers and body recovery bags.
9. Camera for land shots and, if there is enough visibility, an underwater camera.
10. Radios for communication.
11. Diver-down flag.
12. Duct tape.
13. Emergency oxygen and a first aid kit.
14. Contact information for outside resources.
15. Rescue rope throw bags.
16. Contingency pony bottles and main cylinders with regulators.
17. Ear wash and mouthwash solutions.
18. Sunscreen and insect repellant.
19. Potable water.

Witness Statement Report

Interviewer: ______________________________
Department: ______________________________
Date: ______________ Time: ______________
Location: ______________________________
Circumstances: ______________________________
____

Witness

Name: ______________________________
Address: ______________________________
____
Telephone: Home: ( ) ______________ Work: ( )
____
Age: ______ Race: ______ Gender: ______ Hair: ______ Eyes: ______
Height: ______ Weight: ______

Victim

Name: ______________________________
Address: ______________________________
____
Telephone: Home: ( ) ______________ Work: ( )______________
Age: ______ Race: ______ Gender: ______ Hair: ______ Eyes: ______
Height: ______ Weight: ______

Clothing worn: ______________________________
____
Relationship to witness: ______________________________
Known medical history: ______________________________
Drugs or alcohol consumed? ______________________________
Estimated time of immersion: ______

Witness statement

____
____
____
____
____

(Use back, if necessary.)

TENDER CHECK OF DIVER

1. Air on.
2. Hood in place, not under mask skirt.
3. Mask ready, gloves on.
4. Primary regulator: exhale first, breathe three times, check tie-wrap and mouthpiece.
5. Pony regulator, per no. 4. Hose under arm, secured with cover and holder.
6. Dry suit hose on and functional.
7. Diver, press inflator button three times while tender checks pressure gauge. Gauges secured, all hoses trim to body and secured.
8. Diver, without looking, find first shears, second shears, and third tool.
9. Diver, find carabiner at harness tether point. Tether line attached.
10. Diver, find weight belt release. Is it a right-hand release? Ten inches extra webbing?
11. Fins ready, with ankle weights, if necessary.
12. Backups have contingency lines.

THE TENDER SHOULD HAVE AND CHECK THE FOLLOWING

1. Appropriate PFD closed properly.
2. Appropriate footwear and exposure protection.
3. Proper tending gloves.
4. An appropriate time keeper.
5. A water-activated flasher.
6. Eye and sun protection.
7. Back-tethered harness if working from a steep embankment.
8. If acting as a backup tender, a profile slate and writing utensil.

THE WATER SITE MUST HAVE THE FOLLOWING

1. Contingency regulator on full cylinder.
2. Contingency regulator on full pony bottle.
3. If fast water, at least one rescue rope throw bag per diver-tender pair.
4. At minimum, BLS first aid personnel and supplies.

Diver Deployment Record

Location: ____

Date: ____________

Incident Commander: ____

Air temperature: _______ Water temperature: _______

Dive no. _____ Diver: ________________ Tender: ________________

Face acclimated? _______ BP before: _______ BP after: _______

Backup: ________________ BP: _______ Profiler: ________________

90%-ready: ________________ BP: _______

psi out: _______ psi down: _______ psi up: _______ psi in: _______

Total psi used: _______ SAC: _______

Time out: _______ time down: _______ Time up: _______ Time in: _______

Total time down: _______

Average depth: _______ ft. Maximum depth: _______ ft. Visibility: _______ ft.

Weight used: _______ lbs. Heavy/light/okay?: _______

Maximum line out: _____ Line out at finish: _____ Running feet searched: _______

Current: _______ Notes: ____

____

____

Dive no. _____ Diver: ________________ Tender: ________________

Face acclimated? _______ BP before: _______ BP after: _______

Backup: ________________ BP: _______ Profiler: ________________

90%-ready: ________________ BP: _______

psi out: _______ psi down: _______ psi up: _______ psi in: _______

Total psi used: _______ SAC: _______

Time out: _______ time down: _______ Time up: _______ Time in: _______

Total time down: _______

Average depth: _______ ft. Maximum depth: _______ ft. Visibility: _______ ft.

Weight used: _______ lbs. Heavy/light/okay?: _______

Maximum line out: _____ Line out at finish: _____ Running feet searched: _______

Current: _______ Notes: ____

____

____

Line-Pull Signals

Tender-to-Diver Signal

1: Okay? Stop, face line, take up slack.

3: Go to diver's right.

4: Go to diver's left.

2 + 2: Search immediate area.

3 + 3: Stand by to ascend.

4 + 4: Ascend slowly.

Diver-to-Tender Signals

1: Diver is okay.

1: Tender, make note.

2 + 2 + 2: Tangled but okay; alert backup diver.

3 + 3 + 3: Okay, but need help from backup diver.

4 + 4 + 4: Need immediate help.

6 + 6 + 6 (repetitive 6): Found object.

Sample Profile

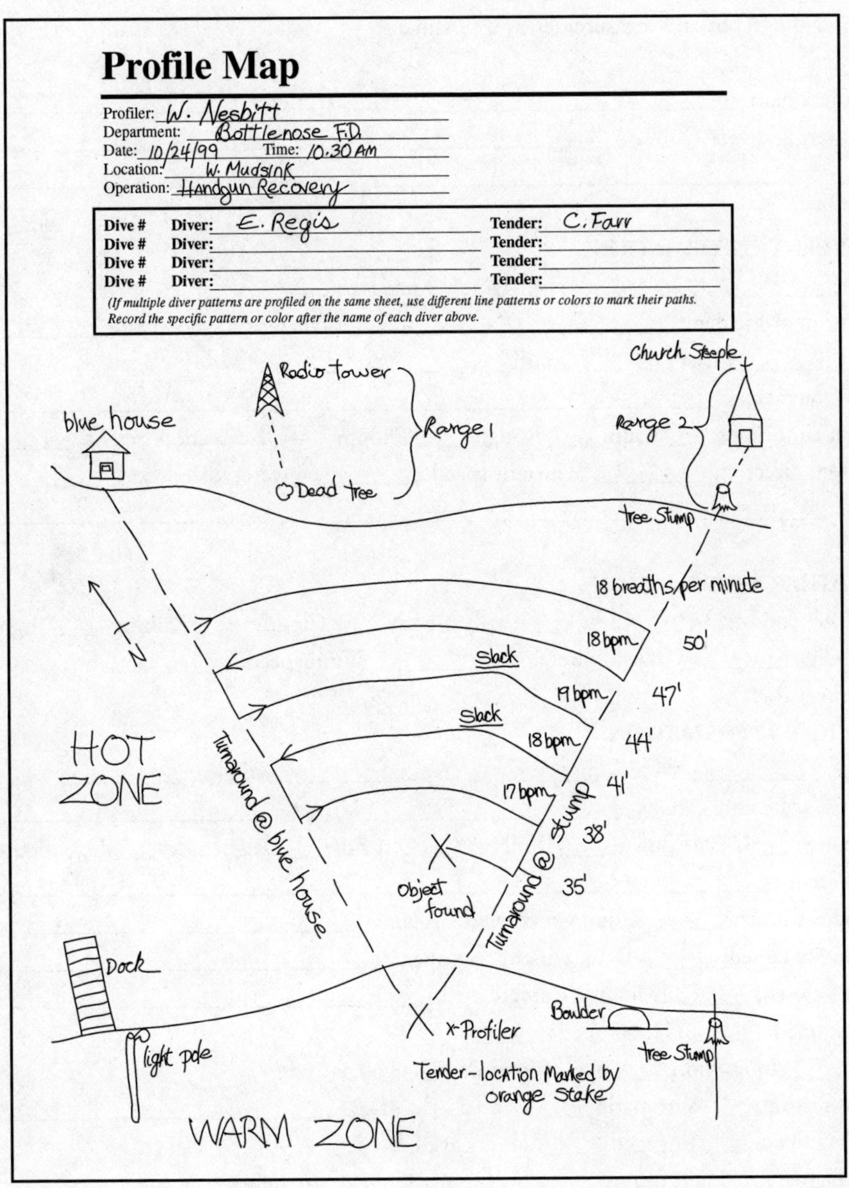

Submerged-Vehicle Checklist

Take photographs and measurements, if possible.

Profiler's name ______________________ Department: __________________
Recovering divers' names: ____
____

Incident Information

Date: __________ Time vehicle found: __________
Location of incident: ____________________ Distance from shoreline: ________
North _____ East _____ South _____ West _____
Shore references: ____
Water conditions: ___ Calm ___ Rough ___ Choppy Wave direction/height: ______
Current direction: _______ Current speed: _______ Water visibility: _______
Other:____

Weather Information

Weather conditions: ___ Clear ___ Partly cloudy ___ Cloudy ___ Raining ___ Night
Visibility: _______ Wind direction: _______ Wind speed: _______

Vehicle Information

Make: ________ Model: ________ Year: ________ Color: ________
License plate and state: ____________________ VIN no.: ________________
Position: ___ Upside down ___ Upright ___ Left side ___ Right side ___ Upended
Comments:____
Windows intact? _____ If no, which were broken? ______________________
Windows closed? _____ If no, which were open? ______________________
Doors closed? _____ If no, which were open? __________________________
Trunk open? _____ Hood open? _____
Keys: ___ In ignition ___ Not in ignition ___ Not found
Transmission: ___ Automatic ___ Manual In gear? ______________________
Parking brake: ___ Engaged ___ Not engaged
Headlights: ___ High beams ___ Low beams ___ Parking lights ___ Off
Windshield wipers: ___ On delay ___ On low ___ On high ___ Off
Damage to vehicle? If yes, describe: ___________________________________
____
____
Objects found and their locations, including victims: ____________________

____
____
____
____

Comments: ____
____
____
____
____

Note: Separate victim checklists must be made for each victim.

Verification signature: ________________________________ Date: __________

RECOVERED-VICTIM CHECKLIST

Take photographs and measurements, if possible. For scuba recoveries, photograph all gauges and equipment as first seen underwater, before removing the victim, if possible.

Profiler's name ______________________________ Department: ________________________
Recovering diver's name: ____________________________________

INCIDENT INFORMATION

Date: _______________ Time victim found: _______________
Location of incident: ___________________________ Distance from shoreline: ___________
North _______ East _______ South _______ West _______
Shore references: ____
Water conditions: ___ Calm ___ Rough ___ Choppy Wave direction/height: _________
Current direction: _________ Current speed: _________ Water visibility: _________
Other:____

WEATHER INFORMATION

Weather conditions: ___ Clear ___ Partly cloudy ___ Cloudy ___ Raining ___ Night
Visibility: _________ Wind direction: _________ Wind speed: _________

VICTIM INFORMATION

Position of victim: ___ Face up ___ Face down ___ Right side ___ Left side ___ Sitting
Comments:____
Marks on victim: ____
Victim's clothing: ___ Bathing suit ___ Wet suit ___ Dry suit ___ Long-sleeved shirt
___ Short-sleeved shirt ___ Long pants ___ Short pants ___ Jacket
Comments: ____
Depth victim found: __________ Entangled?: ______________________________________
Obvious injuries: ____
Comments:____.

DIVER VICTIM INFORMATION

Weight belt on?: _________ If no, describe where found: ___________________________
Amount of weight: ____
Could weights be released quickly if necessary? _______________________________________
Mask in place?: _________ If no, describe where found: ______________________________
Regulator in mouth?: _________ If no, describe where found: _________________________
Tank pressure gauge: _________ Maximum depth gauge: _________

Please write additional comments on back.

Verification signature: ____ Date: ___________

Answers to Study Questions

Chapter 1

The Foundations of a Public Safety Dive Team

1. True.
2. Three years.
3. One that doesn't work under the auspices of a law enforcement agency, fire department, or EMS organization.
4. (1) Procedure, (2) guideline.
5. Yes.
6. Yes.
7. True.
8. True.
9. (1) Five, (2) seven.
10. Record it as part of the dive team's regular operating equipment.

Chapter 2

Dive Team Command

1. Haz mat responses, since water is essentially an alien environment that can kill.
2. Unified command.

3. Awareness.
4. At the command post.
5. Technician- and Operational-level personnel.
6. True.
7. Three to seven.
8. Technician.
9. Never during a public safety underwater operation.
10. True.
11. True.
12. Liaison officer.
13. The diving coordinator and the safety officer.
14. (1) Were the department's SOPs and SOGs followed, and if not, why? (2) Do the procedures and guidelines need amending? (3) Were there enough or too few personnel? (4) Did all of the responding personnel have sufficient training? (5) Were sufficient resources present? (6) Were there any equipment failures or damage? (7) Was the incident properly documented? (8) Did all of the responders behave appropriately? (9) Was all of the equipment properly cleaned and stored for later use? (10) Were all of the divers checked out by EMS, and are they physically and emotionally well?

CHAPTER 3

Dive Team Personnel

1. (1) They must be able to work with one another, and (2) they must be able to work with other teams and associations.
2. True.
3. One thousand feet.
4. One hundred yards.
5. Four hours.
6. Yes.
7. The tenders.
8. Every five minutes.
9. To record the progress of the search by keeping a map showing exactly where the divers have been and what areas have been searched.
10. The backup tender.
11. To search the bottom as directed by the tender.
12. The diver has neither a buddy nor ancillary responsibilities to distract him from the search.

13. The backup diver should be constantly alert to any circumstance that might demand his participation.
14. Because the ninety-percent-ready diver replaces the backup. The backup tender operates with whichever diver makes the descent.
15. Because a stronger diver is more able to meet the demands of an emergency; thus, stronger divers should act as backups and ninety-percent-ready divers.
16. Three divers and two tenders.

Chapter 4

Diver Training

1. Entry-level open-water scuba certification.
2. American-Canadian Underwater Certifications (ACUC), the National Association of Underwater Instructors (NAUI), YMCA Divers, the Professional Association of Diving Instructors (PADI), Scuba Schools International (SSI), and the National Association of Scuba Diving Schools (NASDS).
3. (1) At least fifteen hours, (2) at least thirteen hours.
4. Prospective divers should first practice ascending and descending with a buoyed, weighted line with a knot tied every foot. When they can repeatedly make ascents and descents at the proper rate, they should then practice with their eyes closed. The next step is to practice with straight, unknotted line, and then with no line at all. Finally, they should make ascents and descents while holding a ten-pound weight to simulate a lifeless adult body, which weighs approximately nine to sixteen pounds underwater.
5. Entanglement.
6. By using a blacked-out mask.
7. Diving Technicians, Level I.
8. Because in-house trainers who only train local teams don't gain the experience, knowledge, and skills that professional instructors acquire by working across a wider spectrum.
9. Sixty feet.
10. False.

Chapter 5

Equipment for Shore Personnel

1. Twenty-five feet.
2. Five.
3. By using a line bag.
4. Types III and V.
5. Under the turnout coat.
6. Anytime when working around currents in excess of two knots.
7. (1) As a clean area for divers to stand on while suiting up, and (2) as protection against the elements.
8. 3/8-inch diameter.
9. Braided polypropylene.
10. Knots of any kind will weaken the overall strength of the line, and loops tend to become entangled with anything they find. Both will seriously increase the drag on the line.
11. Hang it freely, with air circulating on all sides of it.
12. Rope bags are easy to use and repack. They offer quick line deployment and help to keep the dive site neat and organized.

Chapter 6

Diving Equipment

1. 50°F.
2. To compensate for the compression of the air in the suit as the diver descends.
3. To reduce heat loss.
4. Because a rappelling harness has a low tether point, which will put the wearer in a vertical position in the water.
5. Contingency line.
6. At least two pairs of shears.
7. Within the golden triangle.
8. Pull it out and away from the body before dropping it.
9. For a right-hand release.
10. Two pounds.
11. True.
12. Thirty-five pounds.
13. Because these units tend to collect sand and mud, leading to frequent malfunctions.

14. 80 ft.^{3}
15. True.
16. 18 ft.^{3} with an independent regulator.
17. Downstream; i.e., they should free-flow in the event of a malfunction.
18. (1) Depth gauge and (2) submersible pressure gauge (SPG).
19. False.
20. True.
21. One hundred feet.
22. 110 feet.
23. True.
24. True.

Chapter 7
Cylinder Safety

1. U.S. Department of Transportation.
2. Corrosion.
3. False.
4. (1) To provide a freshwater washdown for the tanks, (2) to absorb explosive energy in the event of cylinder failure, and (3) to absorb heat generated by the compression of air during the fill.
5. At from 300 to 600 psi per minute (by gauge) in a dry environment.
6. To allow a round-bottomed cylinder to stand on end.
7. SP6688, SP6576, and CTC890.
8. (1) Ninety days, (2) every three years.
9. False.
10. 70°F.
11. 6 psig.
12. No.

Chapter 8
The Initial Response

1. (1) Review past incidents so as to locate the sites where emergencies are likely to occur, (2) examine those sites for any problems or hazards they might pres-

ent, (3) determine what is required to mount operations in those areas, and (4) make all personnel and equipment ready to the task.

2. Sixty to ninety minutes.
3. Because it helps the witnesses develop trust with the interviewer and it lessens the feeling that they're being interrogated.
4. (1) Show me, (2) Tell me, (3) Write it.
5. Because trauma, guilt, overhelpfulness, or other human emotions may distort perceptions or recollections of the conscious mind, or may even drive the witness to be purposely deceitful.
6. The interviewer takes the witness to three separate locations that look out onto the scene and asks him to point to where the incident occurred. The interviewer takes note of the position of the body, and later draws focal lines on a profile map. Often, if the lines drawn from the torso converge, the point of intersection will be the most likely location of the person or object in question.
7. 25 percent.
8. Because they are less apt to lie or create information as adults might.
9. 100.
10. 2 ft./sec.
11. Fifty feet.
12. The depth of the water.
13. A method of determining position by the alignment of distant objects.
14. Risk-to-benefit.
15. (1) Twice, (2) three times.
16. Increased pressure results in an increased solubility of a gas into a fluid.
17. On ascent, pausing at a depth of about fifteen feet for several minutes to help decrease the risk of lung overexpansion injury.

Chapter 9

Deploying the Team

1. (1) Poor mental preparation, (2) poor physical fitness, (3) poor procedures, and (4) being overloaded with tasks.
2. (1) Have you consumed alcohol within the past twelve hours? (2) Do you feel capable of diving now?
3. (1) 110 mmHg, (2) 100 mmHg.
4. The tender.
5. With its valve facing away from the water.

6. By immersing his face in the water and taking three or four comfortable breaths.
7. (1) A contingency regulator on a full cylinder, (2) a contingency regulator on a full pony bottle, (3) at least one rescue rope throw bag, (4) BLS first aid supplies, plus personnel, and (5) potable water.
8. Turn on the air valves.
9. Because a second stage with a missing or damaged diaphragm will deliver air if the purge button is pressed, but it cannot be cleared and cannot be used normally.
10. That the needle doesn't move.

CHAPTER 10
The Search

1. By walking backward.
2. 150 feet.
3. Feetfirst.
4. Pinching the nostrils and closing the mouth tightly while forcing air into the sinuses and through the eustachian tubes so as to equalize pressure in the inner ear.
5. Right.
6. Stop and face the line.
7. Twenty minutes.
8. (1) At least forty-five minutes, (2) at least two hours.
9. Start with your hands in front of you, palms down. Next, sweep them out to each side, then bring them together again. Then, move them forward one hand-length and repeat the sweep.
10. The area should be divided so that the diver can first search on one side of the object, then the other.
11. So that the victim's family can be removed, sheltered, and restrained by law enforcement personnel, if necessary.
12. (1) If there are weeds, trees, vehicles, debris, or other obstacles on the bottom, (2) if the diver shows frequent slack in his tether line, (3) if the diver's breathing rate was too high, indicating that he may have been distracted or stressed, (4) if the diver missed test items that were placed in the search area, (5) if the angle of the tether line shows that a diver has come off the bottom during a low- or no-visibility search, (6) if the diver covered too many linear feet, and (7) if the tender is unable to follow correct tending procedures.

13. (1) The object isn't in the water, (2) you're in the wrong location, or (3) the scope of your search wasn't wide enough.
14. A diver is halfway through his dive when his tank reaches two-thirds of its starting capacity, and he should be out of the water at one-third.
15. Using the BCD power inflator to gain enough buoyancy to bring you to the surface.
16. Since the air in the BCD expands as you move upward, the ascent rate will increase until it becomes too fast to be safe.
17. Thirty feet per minute, or one foot every two seconds.
18. True.

CHAPTER 11

Search Patterns

1. (1) The size of the area, (2) the area from which the tender is working, (3) the available resources, (4) the current, (5) the depth, (6) the object of the search, (7) the probable contour of the bottom, and (8) known or expected obstacles.
2. Two feet.
3. By picking a landmark at both ends of the pattern.
4. About as far as the divers deploy.
5. Two.
6. At minimum, none. The two backups also function as ninety-percent-ready divers for each other.
7. No.
8. The jack-stand technique lends itself well to bottoms that are nearly flat and for searching for small items, such as weapons, evidence, and personal property. It is meant for a methodical search, rather than one covering a large amount of area in a short time. It can also be used perpendicular to a light current (one-fourth to one-half knot) or parallel to a current of up to one knot.
9. MacKin pivot.
10. False.
11. Direct overhead search.
12. Frame search.
13. 600 to 800 linear feet.

Chapter 12
Communication and Line Signals

1. Because they use sound waves to transmit signals through the water, wireless systems are apt to pick up the background noise generated by boat engines, scuba gear, and echo distortions created by underwater objects.
2. To reinforce the use of basic line signals, needed in the event of equipment failure.
3. Stop, face the line, take up the slack, and prepare for a new signal.
4. Search the immediate area.
5. Surface.
6. I'm okay.
7. I found the target.
8. I need help immediately.
9. By touching one hand to the top of his head.
10. Put one hand on his harness carabiner, then concentrate only on relaxing and slowing his breathing rate.
11. The two divers should clasp hands so that the primary can indicate to the backup exactly what is wrong.
12. I'm already on my pony bottle and I need more air.

Chapter 13
Contingency Procedures and Scene Safety

1. To clip on to the primary diver's tether with a contingency line.
2. The backup connects the contingency strap directly into the primary's harness carabiner and disconnects the primary's tether line. The two divers then make a direct ascent to the surface together, both tethered by the backup diver's line.
3. On discovery that a primary is disconnected, the backup should be sent out beyond the primary's last known location. The backup then descends. The tender then brings the backup toward the primary with short sweeps. He shouldn't try to bring the backup directly into the primary diver's bubbles, since the bubbles may not be directly overhead the primary. Also, if the backup misses his target, precious time will likely be wasted with shotgunning attempts. Once the two sets of bubbles are in the same place, the divers should be together.
4. No.
5. Because unconscious divers can't hold their breath.

6. Because pulling any faster might cause the primary's head to submerge.
7. Ask for identification and contact information.
8. Alpha flag.
9. To forestall reporters from speculating that the team has given up.
10. Thirty minutes.
11. Fifty.
12. To allow static electricity to discharge.

CHAPTER 14
Physiological Components of Diving

1. A body core temperature of less than 95°F.
2. Twenty-five times faster.
3. 3,200 times.
4. Skin temperature.
5. Because their peripheral blood vessels constrict earlier, shunting more blood to the core.
6. Because the peripheral blood vessels dilate, thereby increasing the flow of blood near the skin and in the limbs.
7. True.
8. Bobbing and huddling.
9. Rapid cooling will cause shivering, which will only elevate body temperature.
10. Cool the patient in any manner and as rapidly as possible. If possible, place wrapped ice packs under each armpit, behind the knees, on the groin, the wrists and ankles, and on either side of the neck. Treat the patient for shock, administer high-flow oxygen, and transport him immediately.
11. Dehydration.

CHAPTER 15
Diving Maladies

1. Holding your breath for any reason during an ascent while on scuba or surface-supplied air.
2. The volume of a parcel of gas varies inversely with the pressure exerted on it.
3. Because the skin diver holds his breath for the entire length of the dive, his lungs will compress as he descends. His lungs will then return to their normal

size, and the air within them to its normal volume and density, when he returns to the surface.

4. Because with each inhalation, a scuba diver fills his lungs with compressed air to keep them the same size that they were at the surface, meaning that he must breathe increasingly denser air as he descends. If a diver holds his breath while ascending, the denser air that he used below will expand the lungs beyond their normal elastic capability.
5. (1) While equalizing, (2) while dealing with mask-related problems, (3) while coughing, sneezing, hiccuping, or vomiting, (4) while working hard or concentrating, (5) while managing buoyancy, (6) when overweighted, (7) if he accidentally drops his weight belt, (8) while carrying a body to the surface, (9) while riding a lift bag, (10) while being pulled off the bottom, (11) during underwater tow-sled operations, (12) while hanging on to lines in rough water, (13) while breathing heavily in response to finding the target, and (14) when there are predisposing physiological factors.
6. Any clot, bubble, or clump that is lodged in and blocking a blood vessel.
7. Heart attack or stroke.
8. Pneumothorax.
9. Subcutaneous emphysema.
10. The process by which a diver absorbs nitrogen from the air he's breathing into his tissues.
11. If the pressure is released from a diver's body too quickly, as during a rapid ascent, the transition of nitrogen from a dissolved state to a gaseous state won't occur strictly within the lungs. Rather, nitrogen bubbles will form within the tissues and blood of the diver.
12. True.
13. Neurological or cardiorespiratory symptoms.
14. False.
15. (1) Ascending too rapidly, (2) diving too deep for too long, (3) too many repetitive dives and diving too many days in a row, (4) dehydration, (5) cold stress, (6) exertion underwater and high air consumption, (7) altitude changes before or after diving, (8) postdive skin diving, (9) alcohol, (10) taking a hot shower or bath after a dive, (11) fatigue, (12) obesity and poor health, and (13) patent foramen ovale.
16. Decisions, not diagnoses.
17. Field neurological evaluation.
18. The increased pressure of the chamber forces more oxygen into the diver's body, and it also forces gaseous nitrogen back into solution, thereby reducing the damage and blockages caused by the bubbles.
19. The administration of oxygen.
20. True.

21. True.
22. To avoid carotid sinus reflex.
23. Rupture of the eardrum.
24. Sinus barotrauma.

CHAPTER 16
Contaminated Water

1. By culturing water samples for several days.
2. (1) It's more difficult to find a contaminant without knowing beforehand which chemical you're looking for, (2) the equipment required for analysis is expensive and highly specialized, and (3) it's sometimes more difficult to filter enough water to get a detectable concentration of the contaminant.
3. (1) Mechanical, (2) dilution, (3) absorption, (4) degradation, (5) isolation and disposal, and (6) disinfection.
4. Dilution.
5. Immediately on exiting the water.
6. When he has been cleared to remove them by the decontamination officer or his designee; i.e., only after he has been thoroughly scrubbed and rinsed.
7. True.
8. True.
9. Because the salt crystals will retain moisture in the ear.

CHAPTER 17
Handling the Drowning Victim

1. Because the drowning victim no longer has the ability to scream, raise his arms or head out of the water, or struggle about.
2. Sixty seconds.
3. Ten to twenty percent.
4. A reflexive response that seals off the airway.
5. (1) The freer gas exchange within relatively dry lungs, (2) blood pooling in the core, and (3) delayed cellular death.
6. A period of involuntary gasping underwater that can last several minutes, as experienced by wet-drowning victims.
7. Face down and horizontal.

8. False.
9. One of the straps should pass midway between the victim's elbows and shoulders, and the other should pass midway along this thigh.
10. Treat for hypothermia.
11. Laryngospasm may have trapped enough air in his lungs to keep him buoyant for a while.
12. In the digestive system.
13. Literally "human eating," anthropophagy is the consumption of a human corpse by scavengers.
14. False.
15. Spine up, with the limbs hanging down.
16. Putrefaction of the extremities.

Chapter 18
Submerged Vehicles

1. Two to three minutes.
2. In water more than fourteen feet deep, sinking vehicles landed on the nose and then tended to turn turtle, falling over onto the roof.
3. Roll down the windows and escape through them.
4. The implosion of the windshield.
5. That he has reached the vehicle within approximately ten minutes of submergence.
6. Go over, never go around.
7. From below.
8. Full-face masks equipped with transfer blocks, vulcanized rubber dry suits, and Kevlar® glove liners.
9. (1) The nonbuoyancy of the tool, (2) the difficulty in establishing good leverage in an underwater environment, and (3) not being able to see the operation of the tool in conditions of restricted visibility.
10. (1) After a descent line has been secured to the vehicle, (2) if the divers are using an underwater communications system, (3) if the visibility is over eight to ten feet, and (4) if the sediment is of a type that won't obscure visibility if it gets stirred up.
11. Once the vehicle has been located.
12. False.
13. The Federal Aviation Administration (FAA) and the National Transportation Safety Board (NTSB).
14. (1) Dead lifting, (2) internal buoyancy lifts, (3) pumping out, and (4) attached buoyancy lifts.

Chapter 19
Boat Operations

1. 125 feet.
2. An outboard-powered inflatable.
3. Their inability to carry much weight.
4. By means of a scoop that drops over the water jet, directing the flow toward the bow.
5. At least six.
6. On a small plate affixed to the transom.
7. False.
8. True.
9. Down the centerline.
10. When the boat is abeam the next buoy marking the long side of the rectangle (Buoy 2).
11. Fins and possibly the full-face mask.

Chapter 20
Law Enforcement Operations

1. Homicides.
2. Accidents.
3. Photography or videography.
4. A complete descriptive account that encompasses information from the dive log and a court-ready profile map, along with the divers' interpretations of the whole operation. Once complete, the narrative is turned over to the investigator and becomes part of the case file.
5. True.
6. Because they might correspond with clues in the clothing, skin, and under the fingernails of the victim.
7. Because overweighted divers may accidentally push small items deep into the sediment as they drag across the bottom, or they may accidentally brush a small object into an area that has already been searched, causing it to be missed.
8. Kerosene.
9. Because plastic bags and other flexible containers will touch and rub against the item, marring fingerprints and other important markings.
10. The vehicle should be brought out before the search.

11. By collecting it in a sealed container with a sample of the bottom sediment included.
12. If it is impossible to bag the entire body at depth, at least bag and secure the hands, feet, and perhaps the head, making sure to cover the mouth and nostrils to preserve any evidence that could be trapped there.
13. The case number; the location of the evidence at the scene; the depth at which it was found; a description of the item; the date and time it was located; the name of the diver who located the item; and to whom the evidence was released at the surface.
14. True.

CHAPTER 21

Special Considerations and Technology

1. From overhead and not directed away from shore.
2. No.
3. True.
4. A method of searching in a current while secured to a tether line of fixed length that is anchored to a diving platform that can be maneuvered upstream, downstream, and across the search area.
5. False.
6. About four or five pounds.
7. 45°F.
8. True.
9. Along the upper portions of the weeds.
10. Thirty-three feet.
11. 1,000 feet MSL.
12. Ski harness or Y-shaped yoke.
13. Two and a half knots.
14. Decompression sickness and lung overexpansion injury induced by rapid altitude changes.
15. Attention span.
16. The United States Coast Guard.
17. One hundred feet.
18. Four.
19. Selective availability.
20. Three meters.

21. An image as seen from above. The object will appear well defined and lit on the side facing the tow fish and shadowed on the opposite side.
22. (1) Conducting a rapid search of a large area, and (2) searching deep water.
23. One mode is passive, in which the unit functions only as a receiver capable of locating underwater acoustic beacons. In the active mode, the unit sends out an acoustic ping, and the return signal is then electronically interpreted for either an aural tone or a visual display.
24. They are nonselective, in that they detect all kinds of metals.

Index

C

D

E

F

G

H

I

J

K

L

M

N

O

P

S